An Introduction to Environmental Energy Resources

An Introduction to Environmental Energy Resources

S.R. Manjunatha

CYBER TECH PUBLICATIONS
4264/3, Ansari Road, Daryaganj, New Delhi-110002 (India)
Ph.: 011-23244078, 011-43559448 Fax: 011-23280028
E-Mail: cyberpublicationsdelhi@yahoo.com
Website: www.cybertechpublications.com

AN INTRODUCTION TO ENVIRONMENTAL ENERGY RESOURCES

S.R. Manjunatha

First Edition 2010

Asian Offset Printers

Published by :

G.S. Rawat for **Cyber Tech Publications**
4264/3, Ansari Road, Daryaganj, New Delhi-110002 (India)
Ph.: 011-23244078, 011-43559448 Fax: 011-23280028
E-Mail: cyberpublicationsdelhi@yahoo.com
Website: www.cybertechpublications.com

Preface

The Environmental Energy Resources focuses on the emerging scientific and policy issues at the crux of sustainable development. Our goal is for these reviews to provide scholars, practitioners, and policy makers in a range of fields with a thorough and fundamental analysis of the most critical and integrative issues and controversies in the environment and resource arena. The reviews collected here survey and analyze emerging literature in fast moving and diverse fields. The chapters in the first major section—life support systems—reflect a second generation of critical global science issues.

The authors explore the state of the world's genetic plant resources—how many do we have, and how are they accessed and used. They conclude that although conservation strategies have been successful, these collections must now be secured and made accessible to a much broader range of users. Horvath discusses the state of knowledge in the area of current and advanced construction materials. He employs life-cycle methods to evaluate the relative economic and environmental costs and benefits of a range of materials, which are becoming increasingly important as the green building and sustainable architecture sectors rapidly expand. Bridge follows with a comprehensive review of mining that takes a social theory approach to a range of environment and sustainability issues, presenting a picture of an evolution, sometimes slow, toward better environmental management capacity and cleaner production techniques. His review evaluates institutional changes that would facilitate the capture of benefits for communities through processes that also lead to increasing corporate, social, and environmental responsibility.

—Author

CONTENTS

1

Introduction

ENVIRONMENTAL IMPLICATIONS OF THE ENERGY LADDER IN RURAL INDIA

The rural population in India depends on traditional biomass fuels (wood, animal dung and crop residues) for meeting their energy needs. Though nationally there has been a decrease in the share of traditional fuels, as biomass is being increasingly substituted by commercial fuels, the shift to commercial fuels has been negligible in rural areas. In these areas, the monthly consumption of fuelwood has increased per person from 16.24 kg in 1987-88 to 17.27 kg in 1993-94 (NSSO 1997).

Energy consumption in rural areas is determined by: the availability and accessibility of local biomass resources; the competing uses for biomass; availability of commercial energy sources; the cost of commercial fuels such as LPG and kerosene (NCAER 1985); absence of an effective supply especially in the rural areas (TERI 1998); and inadequate production and refining capacity of LPG (Ravindranath and Ramakrishna 1997).

One reason is the range of policies on supply and pricing of commercial fuels, which discriminates between urban and rural areas. The Government of India gives a subsidy on kerosene and LPG, which mainly benefits the urban population, where these fuels are easily available. The price paid by the consumer for kerosene is 69% of the production and supply cost, for LPG it is 61% and for electricity it is 48% (TERI 1992). For commercial fuels, the supply and distribution, as well as the prices, are regulated by government agencies. Kerosene is supplied through the Public Distribution System (PDS) - a network of fair price government shops. On the other hand, fuel-wood is neither subsidized nor is its distribution and supply controlled by the government. As a result, rural consumers often end up paying much more than the controlled rates fixed by the government (TERI 1996, Sharma and Bhatia 1986).

Figure 1 gives the energy ladder in rural India, and barriers inhibiting transition to commercial fuels for cooking. Factors that can bring about transition to commercial fuels are highlighted in the boxes, and described in detail in the following sections.

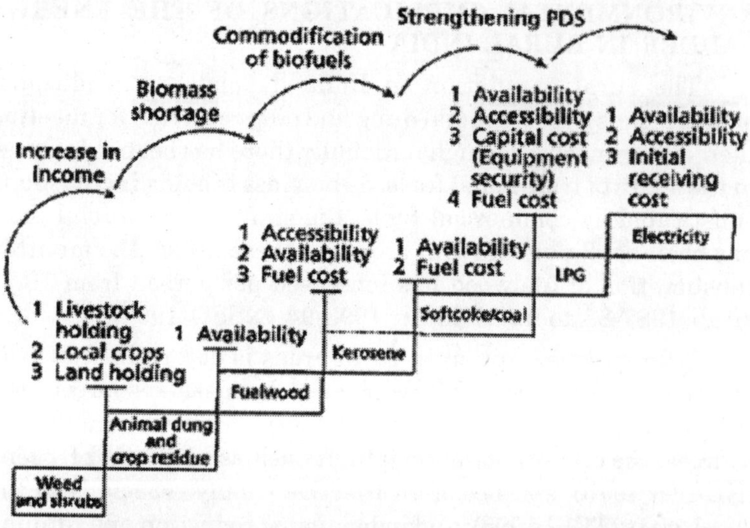

Figure 1: Energy ladder in rural India

ENVIRONMENTAL IMPACTS OF INCREASED BIOMASS USAGE

Biomass resources are at present being over-exploited. There is evidence that where the land is already degraded, fuelwood extraction can make things worse (TERI 1996). Deforestation and burning of biomass significantly contribute to greenhouse gas emissions. In India it is estimated that of the 68.3 million tonnes of carbon released annually due to biomass burning, fuelwood accounts for 82.3% (Kaul 1993). In addition to carbon dioxide emissions, wood burning also creates the products of incomplete combustion (PICs) which have a global warming potential as great as carbon dioxide itself (Smith 1992).

Using agricultural residues and dung as energy sources instead of using them as fertilizers can reduce the soil nutrient level. Out of the total nitrogen content of dung production of 1.35 million tonnes annually, 0.4 million tonnes is lost through the burning of dung as fuel (Ravindranath and Hall 1995). This can have economic implications for local communities as they are forced to buy food if their food requirements are not met from the household plots, due to decrease in productivity of the land (Agarwal et al 1997). Over 136 million tonnes of crop residues are burnt as a fuel in the domestic sector (TIFAC 1992). This results in a significant loss of nutrients to the soil. The water retention of the soil is also reduced increasing the need for irrigation. Crop residues not only maintain the organic content and humus of the soil, but also provide surface protection. Thus a reduction in crop residues makes the soil vulnerable to erosion and drought.

Health Impacts

At the local level, receding forests add to the drudgery of women who have to travel longer distances in search of fuel or, in extreme situations, are forced to switch to inferior fuels such as roots, weeds, leaves, etc. An estimated 84% of rural women aged 10-59 years are affected by fuel-wood scarcity in India (UN 1995). Inefficient combustion of biofuels in traditional cookstoves produces smoke which can cause a variety of health problems such as conjunctivitis, acute respiratory infections, upper respiratory irritation, etc. (Batliwala 1995).

HOUSEHOLD RURAL RENEWABLE ENERGY PROGRAMMES

The Government of India has launched a number of household energy programmes to check deforestation, provide a clean and cheap source of energy, improve soil quality, reduce indoor air pollution, and reduce the drudgery of women, especially in rural areas. Among these, the National Project on

Biogas Development (NPBD) is the largest investment programme and the National Program on Improved Chulhas (NPIC) has disseminated the largest number of stoves. The programmes were launched in the early 1980s and so far about 2.7 million family size biogas plants (capacity of 2cu.m) and 28 million improved cookstoves have been installed in the country (MNES 1998). Wherever the programmes have been successful, local people have benefited from savings in fuel and time, reduced smoke levels, health improvements, cleanliness and hygiene, etc.

BENEFITS OF HOUSEHOLD ENERGY PROGRAMMES

Fuelwood Conservation

In areas where fuelwood is scarce and commercial fuels unavailable, technologies have made a significant impact on the local biomass resource base. In the western Ghat (hilly) region of Karnatka in south India an annual saving of 588 tonnes of wood was achieved from the 2000 stoves installed, making a significant impact on local biomass resources (Ravindranath and Hall 1995). Biogas has reduced the dependence of local communities on protected areas for meeting their fuelwood requirements. In the village of Dedikayal, in Junagadh district of Gujarat, a successful biogas programme has reduced the annual inflow of 300 cartloads of wood from the forests to 15-20 cartloads (Dutta et al 1997).

Improvement in Soil Quality

The use of biogas slurry as organic manure can lead to substantial improvements in soil quality. At the national level, the plants are estimated to be generating about 41 million tonnes of enriched organic manure per annum (MNES 1998). The loss of nitrogen in aerobic decomposition from conventional composting is avoided for the 10 million tonnes of dung used in biogas plants (Ravindranath and Hall 1995). In village of Pingot in Gujarat, the use of wet biogas slurry reduced the weed intensity, and the crop yield was further increased by using a mixture of chemical fertilizer and slurry compost (TERI 1995).

Reduction in Indoor Air Pollution Levels

Reductions in smoke, and in the incidence of eye and lung diseases are perceived as major benefits by over 35% of the total chulha (stove) owners in the country (NCAER 1993). Likewise, in an assessment of impact of biogas, 38% of women felt that its use has reduced the level of smoke, and contributed towards cleaner and more comfortable kitchens (Dutta et al 1997).

Other Benefits

Studies show that in households using biogas for cooking (Figure 3), the time spent by women in cooking, making cow dung cakes, collecting fuelwood and washing blackened utensils has come down, giving them more time to attend to their childrens' needs (Bilgi 1997). Economic benefits in terms of reduction of labour input, diversion of female labour to productive activities, fuel purchased, etc. have also been reported (Mittal 1997, Ravindranth and Hall 1995).

The transition to commercial fuels is difficult because of a weak distribution system, inequities in distribution and high costs, especially when traditional fuels are available at zero cost. There is a need to strengthen the Public Distribution System to prevent pilferage and diversion of kerosene for other uses. However, the household sector must compete with other sectors for commercial fuels, and given the large import bill that the country pays for petroleum products, the rural sector is unlikely to be given preference (TERI 1996).

As biomass becomes increasing scarce and the environmental effects become more obvious, people will have to start paying for it in rural areas (Leach 1992). This may well increase the demand for commercial fuels, leading to in improvements in the distribution network. National level surveys have indicated that people will move to commercial fuels as they become more wealthy (NSSO 1997). On the other hand, some case studies have found that as household income goes up, the quantity of biofuels used also increases (Singh 1993). Household energy technologies such as improved chulhas and biogas need to be promoted more strongly to ensure a sustainable use of the local resources and an improved quality of life for the rural people.

IMPROVING THE ENVIRONMENT CAN LEAD TO BENEFITS FOR HOUSEHOLD ENERGY THE ENVIRONMENTAL DILEMMA

In the past two decades, environmental issues have become an ever-increasing global concern. Many projects will aim to reduce the consumption of local natural resources, especially forest resources. Increasingly, household energy projects are addressing indoor air pollution to reduce health impacts on women and children. Global issues are now also influencing energy strategies in developing countries.

The Issues

There are a number of possible impacts of household energy on the environment, which will affect both the local and the global environment. Some of the main concerns are as follows;

Resource Depletion

Fuelwood depletion is the problem that can have the most immediate impact on the community who use the forest resources. It is now widely accepted that fuel use is not the main cause of deforestation, as compared to the impact of commercial logging and land clearance for agriculture. However, where forests are depleted, fuel demands add increased pressures on remaining forest areas.

As new reserves of coal, oil and natural gas are found around the world, fossil fuel resource depletion is no longer considered the major crisis in the global energy sector. Nonetheless, the need for conservation for future generations should still not be forgotten. Fossil fuel consumption is also directly related to global warming (see below).

LOSS OF BIODIVERSITY

Loss of forestry has a direct impact on the natural surrounding environment. Some of the papers in this issue discuss the knock-on effects of deforestation, which include loss of forests environment, soil erosion and local climatic changes.

All of these will add to the loss of the diversity in the local plant and wild life.

Indoor Air Pollution (health)

As Boiling Point 40 addressed in detail, the greatest impact of household energy on human health is due to pollution of the indoor environment. Smoke in the kitchen has been shown to be one of the main factors in many deaths in children under the age of five, as well as the cause of many illnesses and discomforts for women.

Global Warming

Global warming has been shown to be caused by two factors.

- the release of so-called greenhouse gases to the atmosphere; carbon dioxide (CO_2), which is released when fossil and biomass fuels are burned, and to a lesser extent, methane (CH_4) and nitrous oxide (N_2O), formed when there is incomplete combustion of fuels.
- Deforestation: trees and plants absorb CO_2 so when the forests are destroyed, less CO_2 is removed from the atmosphere, so increasing the greenhouse effect.

Although not seen on a daily basis, over a number of years the impacts of global warming will be felt in different parts of the world. Some authorities claim that recent natural disasters around the world have been due partly to global warming whose main effect is to raise global temperatures, thereby increasing drought in some places, and creating floods in others. A range of other climatic changes are also predicted.

Win-win Benefits

It would be easy to view environmental conservation as a costly addition in time and resources to a household energy project. A win-win benefit is one in which a benefit in one way also gives a benefit in another way. In fact environmental gain will, in most cases, result in additional wider benefits. This can be classed as a win-win situation, where both the environment and the communities involved benefit. A few examples of win-

win benefits are given below. Improved stoves reduce both the time used in fuel collection and the amount of fuelwood needed

Improved cooking stoves should give a higher efficiency, which mean less time collecting wood and cooking, so more time can be devoted to the family or to productive enterprises. The environmental gains are of course, reduction in fuelwood consumed, causing less impact on forestry and lower global warming emissions.

REDUCED SMOKE INCREASES BENEFITS TO HEALTH

Another benefit of improved cooking stoves and the use of other higher grade fuels are the health benefits. By reducing the use of fuel wood, there is a significant reduction in exposure to smoke emissions for the cook and for her children.

IMPROVED FOREST COVER IMPROVES WATER RETENTION AND NUTRITION IN THE SOIL

There are a wide range of benefits which increased forestry cover has on the local ecosystem and the knock-on benefits to farming. These include improved water retention and soil nutrition. In addition, properly managed forests will allow for a wide range of biodiversity, therefore animal and plant species will not be lost.

USING URBAN WASTE AS A FUEL REDUCES ENVIRONMENTAL HEALTH HAZARDS

Using urban waste as an energy source is a new concept in the household energy sector. The workshop described by Dunnett details the win-win opportunities for both energy supply and for reducing the potential environmental hazards of rotting waste on urban streets.

NEW LIVELIHOODS ARE GAINED FROM ENVIRONMENTALLY SOUND TECHNOLOGIES

The introduction of improved and renewable energy technologies often opens up opportunities for developing local business. A commercial approach to producing, distributing and maintaining improved stoves has been shown to be the best and most sustainable approach. The article by Kimani recommends a commercialized approach even in refugee situations, where livelihoods are in a very fine balance.

Obstacles

However, there remain a number of obstacles for the introduction of environmental conservation measures in the household energy sector.

Land ownership and forestry rights: this is addressed in the article by Klunne, where lands used for fuel collection are being returned to large land owners for large sugar production.

Access to fuels: The article by Malhotra describes the difficulty that rural populations in India have in accessing kerosene and LPG, due to subsidies which favour the urban communities. Limited access to alternatives to fuelwood can often intensify the pressure on local forestry resources.

Access to improved technologies: The main barrier to accessing improved or renewable energy technologies is the initial cost. Even if the day-to-day cost of the energy supply is cheaper, the cost of the technology may be too great for many people in poor communities.

Lack of baseline data on environment and socio-economic conditions: Lack of evidence or data on environmental problems will often delay action from governments or concerned organizations to mitigate the problems.

Economic conditions of target group: While the affected group may be fully aware of the erosion of their environment, and the increasing difficulties in accessing fuels, many are not in a position to address the problem due to lack of economic stability.

Possible Solutions

Below is a brief list of some possible solutions to the environmental dilemmas of household energy:

- Reforestation and forest management creating renewable forestry resources
- Increased efficiency of combustion technologies

- Optimum use of waste materials as a fuel, while ensuring higher combustion efficiency
- Implementation of renewable technologies, such as biogas for cooking or solar power for water heating and lighting
- Appropriate financing options, to help poor communities access improved and renewable technologies and fuels
- Appropriate government policies, which take a win-win approach to both the environment and the provision of household energy.

HOUSEHOLD ENERGY AND ENVIRONMENTAL REHABILITATION; OPPORTUNITIES AND CHALLENGES

When a country agrees to host refugees, it becomes responsible for the environmental changes that occur because of their stay. Environmental degradation has always been associated with refugee situations, particularly in sub-Saharan African countries, but although some degradation is inevitable, planning and developing appropriate ways of dealing with the problems can yield real benefits. Although there has been considerable experience with environmental issues in refugee situations worldwide, lessons are ignored and the same problems seem to be confronted whenever there is a new refugee situation. Also, there is a tendency to separate environmental aspects from other parts of refugee assistance, such as food and shelter, and experiences gained in other assistance fields is lost.

The supply of fuels, promotion of energy saving devices and energy saving methods, are the main strategies adopted in many refugee situations. Other possibilities include introducing regulations to control fuelwood collection, training and public awareness on fuel use, always within the context of the social cultural, economic, political considerations on the ground.

It is now clear that refugee situations are not temporary, so plans should involve the mid-term to long-term effects, but nevertheless be able to adapt to rapidly changing refugee situations. For example, a project set up to look at local deforestation may find complications such as:

- both the refugees and the local population may be living in extreme poverty;
- the host government might not be able to support a development or poverty alleviation project
- there could be general insecurity in the areas in question.

In such circumstances, offering relief may not be enough and agencies may need to act as mediators and try to alleviate local poverty, to ensure that:

- the impact of the displaced community on the local environment is minimal;
- work undertaken will cope with possible future environmental problems;
- the security and well being of the displaced community is not put at risk.

The idea that refugee communities should receive free aid without responsibilities must be replaced by the concept that refugee communities are like any other rural, poor population, except they have fewer resources and opportunities.

THE ROLE OF GTZ IN EMERGENCY SITUATIONS

Although GTZ is traditionally a development agency, it is increasing involved in relief and emergency programmes including environmental interventions e.g. natural resources protection, afforestation, supply of firewood, firewood substitution, energy conservation and education measures. By bridging relief and development concepts, it has been possible to reduce the impact on the environment caused by refugee situations.

The Problems

- The traditional approach has been to focus on quick and massive dissemination and adoption of improved cooking stoves and techniques.
- The issue of land-ownership; for example, in tree planting, having control over the ownership of trees, power to decide which trees to plant, which tree products to harvest and sell etc. is vital.

- Household energy programmes have also, in the past, been mistakenly approached from a purely technical perspective, forgetting that every situation has its own peculiar characteristics.
- It is vital for those organizing household energy programmes to have adequate technical awareness to evaluate the benefits of various interventions.
- High staff turnover, prevalent in refugee situations, has meant that unless there is proper documentation, in-depth knowledge of project interventions and experiences is lost.
- Integrating household energy activities into the general refugee operation has often been neglected.
- Lack of participation and adequate representation of those involved in household energy programmes has often occurred because the lack of time and the emergency nature of the situations.

Principles that should be Adopted

The following guiding principles need to be taken into account when formulating household energy programmes. Any interventions should:

- recognize the emergency nature of the situations but appreciate that when the emergency phase is over, the displaced communities will remain;
- contain an element of development, where the displaced communities and the local community share the benefits resulting from the interventions.
- interventions should aim to offer 'wider' and 'longer term' solutions e.g. contribute to incomes, improve personal security, reduce workloads etc
- be fairly simple and adaptable; any cash or material inputs should be minimal. Emphasis should be given to the use of locally available resources
- be flexible and easily replicable taking advantage of existing skills; they need to be exportable and useable in most places

- should not be considered 'free hand-outs' similar to the food and non-food items freely issued to the community.

CONVENTIONAL VERSUS IMPROVED (FIREWOOD) STOVES

If wet or green firewood or small twigs are more available than dry wood, they perform better with three stone fires unless improved stoves are specifically designed for them. which in most cases they are not. The reason is that green and/or wet firewood need more air, and emit more smoke than dry firewood. In such situations, a shielded open fire might be more appropriate.

On the other hand, when small twigs and brush are used as fuel it is necessary to lower the pot to the fire and/or reduce the amount of air going into the fire. Incidentally, improved stoves have fixed or semi-fixed designs and are not as flexible as the three stones fireplaces. That is why it is common to find a household cooking on a three stone fire next to a functional improved stove.

Appropriate Interventions

A comprehensive 'package' covering all aspects of household energy should be considered. For example, stove adoption increased when the household was supplied with firewood only if it constructed and used an improved stove. Until recently, programmes have mistakenly based their success on the number of units given out rather than the number used. Technological solutions should be arrived at by consulting the community and should be part of the general refugee assistance programme. Furthermore, technological interventions should encourage and support the community to look at their lives as dynamic and progressive and contribute to (self)-confidence building.

Factors to consider when developing technologies in household energy programmes:

- Familiarity - technologies should be based on what the community have and are used to;
- Simplicity - they should not add extra burdens to the community in terms of time, labour or education and

should take into account the prevailing gender considerations;

- Affordability - the basic models should not necessarily require cash investments;
- Durability - they should be strong e.g. not be destroyed by rain and should be built and repaired with minimal outside help
- Sustainability - they should be built, maintained and reproduced without depending on agencies to provide external inputs like clay, metal etc.
- Flexibility - it should be possible for them to be dismantled, built up again and modified without outside help
- Adaptability - they should encourage users to further improve them e.g. with mud/clay lining, additional wind breaks, lower pot rests, etc.

Ensuring Continuity of Skills

In many displaced communities situations, the big international agencies leave after the initial emergency phase is over. It is important for the original agencies to ensure that they pass on their skills to smaller NGOs during their stay to enable them to carry out the programmes, through secondment of trainers, or through exchange visits to similar projects elsewhere. A major goal for agencies should be to train skilled people among the displaced and local community to carry out household energy conservation programmes once the outside agency goes away.

COMMUNITY - STAKEHOLDER - PARTICIPATION IN HOUSEHOLD ENERGY PROGRAMMES

Communities should be actively involved in the planning, implementation and evaluation of household energy programmes. This is because the communities have the greatest interest in the well being of their immediate environment. Lack of participation results in hostilities, suspicions and indifference. It is important that the stakeholders themselves decide on their level of involvement.

COMMERCIALIZATION AND INCOME GENERATION

For household energy programmes to be sustainable, they need to be commercialized. Initial subsidies are necessary in the research, development and awareness creation phases but have to be carefully applied and phased out as soon as possible. Opportunities for the majority of displaced communities to earn incomes are usually limited. The argument that the people do not need money because their needs are catered for is incorrect. In fact, it is difficult to imagine a situation where even the basic needs of such communities are adequately met. The recommended strategy for on-site technology production is to support private entrepreneurs or well-organized groups. External support should be limited to initial support covering:

- technical training;
- marketing and book-keeping;
- skills development;
- provision of working tools on a credit or revolving fund basis.

IMPROVED COOKING SYSTEM PROMOTION IN GOMA, DEMOCRATIC REPUBLIC OF CONGO (DRC)

This approach was developed by GTZ for the Kahindo Refugee Camp (DRC) with a population of approximately 600 000 people comprising 10 000 households.

- Some days beforehand, a trainer together with the community leader identified three small groups to be trained on a particular day. In every group, one household preferably belonging to a vulnerable family or the community leader was selected as the 'model household'.
- Five families around this household were asked to report to the model household for a two hour training session. Unless the household was seriously handicapped, it was asked to provide the necessary materials for the stove construction demonstration.
- During the demonstration, participating households were required to take part (hands-on) in the actual

construction. The trainers re-visited trained households three days after the training to check on how the households were getting on.

- Key trainers were identified from every block representing several groups. A regular follow-up schedule to the various households was developed, which was easy and cheap since the trainers were refugees and lived among the community.

FUELWOOD DISTRIBUTION AND SUBSTITUTION

It is being increasingly debated whether providing at least the 'minimum' fuel needed in refugee situations should be considered alongside health, food, water, shelter etc.

Fuelwood Distribution of an Organized Supply

Trucking of fuelwood is a very expensive, especially where road networks are not well developed. The fuel should not be harvested and trucked from within walking distances from camps or settlements. Ideally, the fuel should be bought from private, government plantations or individual farmers and be accompanied by a programme of replanting. Fuelwood Collection Support It is likely that some fuel will need to be collected. The communities need to be supported to fetch fuel from distant places in order to preserve their immediate environment. There are several strategies for supporting fuel collection:

- In situations where security is a problem, trucks can be provided to take people to fuel collection points. The truck can then transport back small quantities of fuel for the households unable to collect it, whilst the collectors return in small groups. Security should be provided along the route;
- Alternatively, the community could be asked to organize fuel collection groups. Escorts, either armed police or groups of men could then accompany the collection groups
- If women and children are responsible for collecting the fuel and if they are exposed to physical assault, the

community should be asked to consider ways of involving men in the exercise.

Fuelwood Distribution

Problems are caused by:

- distribution sites that are located far away;
- lack of information about where and when the distribution is taking place;
- physically disabled preventing people from reaching collection points

It is necessary for agencies to get together with the communities to develop fair fuel distribution strategies which will reach everyone. They should also agree on penalties for those who do not keep to the agreed strategies.

Types of Food

The type of the available food basket should be a key consideration in household energy programmes. For example, it is better to supply milled rather than whole grains because the former requires less energy to cook if the society finds them acceptable. Agencies involved in the food supply should be fully linked with household energy programmes.

Ensuring Continuity

Documentation: It is important to document achievements, setbacks and other experiences in household energy interventions. Once this is properly done, the impact, for example, of the frequent staff turn over and short term funding of projects is minimized and sustainability ensured particularly after the major agencies go away.

INTEGRATING ENVIRONMENTAL ISSUES INTO OTHER SECTORS

To maintain the contribution to the security and well-being of the assisted communities, it is important to integrate household energy interventions into other activities. This can be done through the training of the extension workers e.g. community health and development workers. Through their daily interaction with the communities, these extension staff have established rapport, confidence and understand with the community.

COMMUNITY PARTICIPATION - THE ENVIRONMENTAL WORKING GROUPS (EWG)

In the Dadaab refugee complex in Kenya, there is an Environmental Working Group (EWG) which provides a forum for discussing local natural resources. This working group is chaired by UNHCR (United Nations High Commission for Refugees) and has representatives from refugees, local communities, government and other agencies working in the region. As refugees outnumber local people by a ratio of about ten to one in Dadaab, this method of solving problems is especially important as a channel for airing local views and concerns. Because everyone feels fairly represented, it has become the main way for getting things done. Similar working groups have been adopted in other refugee situations under different names.

CONCLUSION

Refugee situations are not temporary, and the planning for household energy and environmental mitigation programmes in general should be considered in the short, medium and long-term. Most household energy approaches have been tried and refined in one situation or another. It is cheaper to borrow and adapt experiences from such projects in new situations. Exchange visits and secondment of staff to similar projects should be actively encouraged. Conventional development approaches can be successfully adapted and applied to refugee situations. The approaches need to be flexible, dynamic and adapt to changing situations, and even look further into repatriation, rehabilitation and re-integration phases.

It is important that qualified and experienced staff be entrusted with the implementation of these approaches. Sound and objective decisions on and from the ground are necessary not only to ensure overall programme success but also to influence thinking at the different levels of decision making.

2

Problems of Poverty and Development in India

As India prepares to enter a new millenium, the degraded state of India's natural environment cannot escape comment or analysis. Some believe that the deterioration in the environment is of such magnitude that all development must cease for the planet to survive. Others dismiss the entire environmental movement as comprised of loony troublemakers who have no right to interefere with the sanctity of private property and private enterprise. Still others berate the environmental movement for being an exclusively middle-class movement that is irrelevant to the class struggle.

Lost in the middle of these antagonistic positions are those who see environmental pollution as a serious issue that affects people cutting across class lines but also recognize the unfairness of the existing order in which the earth's resources and benefits of modern technology are monopolized by a minority of the rich on this planet. For them, issues of environmental degradation must be tackled on a war-footing, but they reject solutions that preserve the unfair distribution of the world's products and natural resources. For them, fairness and justice - the idea that there must be progress for all, cannot be sacrificed in the name of "saving the environment". For them, environmental concerns must be integrated into the general class struggle, and against the tendency of private interests to violate and exploit both people and the environment. Their world-view is in sharp contrast with those environmental elitists who wish to enjoy all the fruits of modern industrial development while denying those same benefits to the proletarian masses who must struggle and survive under the most primitive of human conditions.

Air and Water Pollution in Delhi

Take the issue of air and water pollution in Delhi and the recent supreme-court rulings in that regard. Few can deny that Delhi's air is deadly to breathe. By some estimates, one in three is now afflicted with chronic breathing ailments. Almost a sixth of all children are reputed to suffer from lead induced mental retardation. Gastro-epidemics (resulting from water-borne pollutants) have been all too frequent in recent years, with children being especially vulnerable. Yet, for some industrialists this is not a "serious" enough problem to warrant further regulation or even ensure the implementation of existing regulations. Curiously, some unionists have adopted the absurd position that this is not a "class" issue, as if the poor and the working class don't suffer from lung diseases or water contamination. If anything, it is the other way around. The poor are the first to suffer from the effects of air and water pollution. While the rich can limit their exposure to air-borne toxics by driving around in air-conditioned cars, the poor who must walk to work, or use bicycles or public transport simply can't escape the ill-effects. The rich can buy bottled spring water or distilled water, but the poor usually can't afford anything but tap water. Rather than dismiss such concerns, unionists and others involved in the class-struggle ought to be the first to take up the battle against air and water pollution.

Instead, this aloof attitude has led to solutions that don't go far enough, and are decidedly unfair to the poor. For instance, the Supreme Court has ruled that "polluting industries" must be relocated out of Delhi. Does that mean that these polluting industries are now free to destroy the health of residents in Ghaziabad, Meerut or Rohtak? Does the Indian Supreme Court only care about the health of those who reside in the nation's capital? And what of the hapless employees of those industrial units that were given the orders to relocate? How are they supposed to cope with unemployment and lack of income? Shouldn't the real solution have been to insist upon correcting manufacturing processes so that they didn't pollute in the first place? Of course, one of the obstacles to any environmental regulation is that businesses will scream that it hurts their "bottom line". When the right to profits is supreme there is little arguing with such views. Yet, not many in the environmental movement consider public ownership of industry as a possible solution where environmental regulations could be enforced across the board, without the risk of corruption or lobbying by private business interests.

A significant source of pollution in India's cities are vehicles - cars, two-wheelers, three-wheelers, trucks and buses. Although there have been small steps in moving towards unleaded fuel, little has been done to improve and encourage the use of public transportation. Railways have been neglected and buses for commuters are poorly maintained. Private bus, tempo and truck manufacturers have largely gotten away with producing poor quality vehicles that routinely exceed emission norms. In a socialist economy, private cars would not be a priority. Expanding the rail network and improving the quality of buses and goods transport vehicles would instead, get their due importance. Unfortunately, a majority in the environmental movement are unable to break from their affinity with private ownership and support of private enterprise. As a result, society makes little, or very slow progress in solving these pressing issues.

The Issue of "Big Dams" and National Sovereignty

Another divisive issue for India has been the growing controversy over "big dams". Over the years, critics of India's dams have presented a host of damning evidence pointing to the deplorable neglect of displaced people - especially poor tribal communities. Critics have also correctly pointed out how many dams are started but never completed, or else poorly constructed. Corruption and mismanagement have often led to huge cost overruns, and too often, there is a lack of follow-up in building adequate feeder canals, or else, the canals that are built are not properly maintained.

But rather than calling for the rectification of such grievious errors, many have adopted an idealogical and blanket opposition to all hydro-projects. This has led to grave consequences for the nation. By increasing the nation's reliance on thermal power stations, it has created chaos in the national grid. Thermal power stations are much less suited to handling fluctuating demand. Given the poor quality of India's coal, and the long distances it has to be transported, India's thermal power plants are more polluting, relatively expensive to maintain, and prone to frequent tripping. What is even worse is that the neglect of hydro-power is leading to a growing reliance on even more expensive imported fuel based power plants of the Enron-type. *Not only are these very costly (and also entail environmental risks), they involve a grave loss of national independence.*

While it is not surprising that US-based anti-dam groups might encourage the shift to oil and gas-based energy, it is tragic how some Indian environmentalists are falling into that trap. As it is, India's oil import bill is too large for the country to afford. By building more oil and gas based power plants, India will become even more vulnerable to pressure from the US, which has a virtual stranglehold on the world's oil trade.

It is important to note that historically, manipulating water resources has been a powerful tool for political conquerers. When Mandu was conquered by invaders from the North-West, Bhojpur's dams were destoyed so as to demoralize and weaken the local peasantry and destroy the legitimacy of the Parmar rulers. The Parmar Rajahs of Bhojpur were particularly renowned builders and admired and venerated for their acumen as great architects and civil engineers. By destroying Bhojpur's dams, Mandu's Afghan conquerers were able to hasten their defeat. Several centuries later, India's British colonial rulers, through a policy of deliberate and wanton neglect, destroyed Bengal's traditional water-works, thus compromising the region's ability to feed and defend itself. It is therefore, not surprising that powerful neo-colonial interests in the West would wish to keep India energy deficient, and dependant on imported sources.

The negative reation to India's nuclear tests, and the resistance to India's nuclear power policy, stems at least partially from the desire of some to constrict India's development. Even though India's nuclear power plants use resources that are locally available, and have a safety record that matches the highest international standards, India's nuclear policy has been severely criticized in major international fora. International groups are particularly hypocritical in their condemnation of India's nuclear efforts when in fact, it is the rich industrialized countries like France, Sweden, Japan (and many others) who are the biggest users of nuclear energy. Although new technologies reduce the problem of nuclear waste-management, and make it much more manageable, many in the West prefer to foist oil-based energy use in oil-poor countries like India.

"Independence in the energy sector is central to the national sovereignty and national well-being for an oil importing and developing country like India". This was the view expressed recently by Dr S. K. Chopra, Senior Advisor to the Ministry of Non-Conventional Energy Sources. Dr Chopra was speaking after making a presentation on "Energy Policy for India - Towards Energy Independence" in Delhi. A strong advocate of alternative energy, he spoke against the oil intensification of the Indian economy. He pointed out that there was inadequate use of renewable sources of energy which must be stepped up. *"We are one of the leaders in harnessing wind energy, but must widen our net"*, he remarked. On the atomic energy front, Dr Chopra stated that India and France had begun harnessing atomic power at the same time. But while France was getting 80 per cent of its energy needs from atomic power today, its contribution in India was only 2.35 per cent of the entire need.

Energy Management

Ideally, India should do far more to manage the demand for power. Sections of Indian industry are extremely wasteful in their use of energy. India's homes could be better designed to be cooler in the summer and warmer in the winter. More could be done to strengthen the national power-grid, minimize transmission losses and rebuild and improve outdated power plants. Non-polluting alternative power sources such as wind, solar and tidal power need far more national support and funding. These should be particularly useful in residential use and in electrifying remote and poorly connected villages of the country. But these alternatives cannot replace conventional sources of energy production entirely. It should be noted that the development of "alternative energy resources" presumes highly developed conventional energy sources. *Wind-turbines require manufacturing that would be impossible without energy supplied by conventional power plants. Solar panels also need advanced high-energy industrial manufacturing.* Hence, it would be naive to argue that India could simply abandon the development of conventional energy resources and suddenly switch to alternative sources. Those who advocate 'small decentralized plants' seem unaware of the heavy requirements of modern industrial production.

The transition to alternative sources will take several decades. In order for India to expand it's use of non-polluting alternatives to thermal or oil/gas based power, it will also have to expand the production of conventional energy in that interim period. Moreover, for energy alternatives to became fully viable, considerably more research and seed capital will be necessary. This is unlikely to come from private sources, and will require greatly increased state-funding.

'DECENTRALIZATION' AND 'SMALL IS BEAUTIFUL'

Although 'decentralization' and "small is beautiful" have become very popular slogans, these are fraught with their own contradictions. Consider the proliferation of "small" private generators. Powered by petroleum derivatives - these are energy-inefficient, extremely noisy and highly polluting. But unlike Narmada-Bachao, there is no *Generator-Bachao* movement. Inconsistencies like these could lead to questions about the sincerity of some environmental activists. How many dare to speak out against the growing use of luxury cars, high-powered air-conditioners, and the reliance on private generators?

Another charge levelled against some of India's environmentalists is that they are hostile to progress. That while they are extremely sensitive to the victims of developmental displacement, they are insensitive to the suffering of those who have been denied any benefits of development. In their exclusive advocacy of "small" hydro projects, they are unwilling to consider that these may sometimes be inadequate even in fully meeting irrigation and drinking water needs (let alone generate electricity).

Recently, Gail Omvedt (at Pune University) pointed out how small irrigation projects while helpful were insufficient to meet the needs of some of Maharashtra's small farmers. She made this observation after a ten-year study of districts where small-scale water harvesting projects had been developed to the fullest for over a decade. She also observed how the "small is beautiful" mantra - while initially helpful in improving the lives of poor communities, eventually puts those communities at a disadvantage vis-a-vis richer farmers with access to more advanced irrigation schemes, and allows the rich farmers to continue to politically and economically exploit the poorer farmers.

What is particularly ironic is how some of the "small is beautiful" mantra is posed as a counterpoint to "Western Models" of water management. But, building large dams is not a Western invention at all. India has one of the oldest traditions in elaborate water-management projects going back to the times of Mohenjo-Daro. Many of India's medieval rulers diverted rivers, built artificial lakes, canals, dams and water-reservoirs. Today, the large lakes and reservoirs of Jaipur, Udaipur, Hyderabad or Bhopal may seem "natural" - but they are all the result of human endeavour. The Bhopal lake required diverting a river more than 10 kms away. Neither is silting of rivers exclusively a problem of 'modern development' as some contend. There are references to India's medieval rulers recognizing the problem of silting and organizing campaigns to desilt rivers. Rajni Palme Dutt, in his classic book on colonial rule in India - 'India Today' - bemoans British neglect of the vast irrigation networks that had been constructed and carefully maintained in Bengal and Assam before the Plassey defeat.

As already mentioned, the highly popular Parmar Rajahs of Central India were noted for their devotion to "large" water-management projects. As skilled architects and promoters of large construction projects, they won considerable popular acclaim in their times.

Environmental Destruction in the Past

There is also a common myth that there was no environmental destruction before the onset of the industrial revolution. In fact, some of the greatest deforestation in the lands around the Mediterranean took place in the Greek and Roman era when indiscriminate logging and mining led to rapid clear-cutting and consequent desertification. As the world's population has grown, the tribal practice of slash and burn agriculture has also become less environmentally viable in some places, and in recent times this has been a major obstacle to forest preservation in Africa and even parts of India. Some environmentalists project an idealistic view of ancient practices and tribal customs. But they forget that the lack of modern technology can also be a source of environmental destruction. When the forest cover dwindles, even the cutting of trees by poor peasants for heating and cooking can pose a problem. (Of course, the primary source of deforestation in India is not the construction of dams but indiscriminate commercial logging - which deserves much greater scrutiny and control.)

The history of the planet is marked by environmental destruction and by serious attempts at resisting and managing such trends. Industrialization greatly heightens this conflict. But it would be wrong to romanticize the past. It is important to realize that there is a dialectical relationship between environmental preservation and human technological development. The point is to strike the best and most sustainable balance.

DIALECTICS OF DEVELOPMENT AND ENVIRONMENTAL PRESERVATION

Ultimately, the strength of India's environmental movement will depend on how it builds alliances with popular movements and sympathizes with the concerns of all of India's down-trodden. There is a great deal to be done in terms of cleaning and protecting the environment. But it cannot be done by simply ignoring the poverty that results from under-development. Neither would it be correct to espouse environmental concerns in a manner that preserves the right of the rich to every modern luxury even as the poor huddle in slums in the heat and cold.

Take the "Narmada Bachao Andolan". Activists involved with Narmada Bachao rightly bring up the issue of *"development for whom?"*. Who pays the price, and who garners the benefits of "development" are extremely important issues facing India as a nation. But is the final answer that all development must cease - *or that development must be structured differently?* Is the problem of developmental inequality a cause or more a symptom of existing class relations in society?

At present there is a huge chasm between opponents and proponents of the Narmada dam. Some argue that the dam should have never been built. Others point to how it has already succeeded in preventing the frequently recurring flooding that occurred along the banks during heavy monsoon years. In Gujarat, support for the dam is almost universal and even extends to popular NGOs. The problems of the dam oustees are real and exceedingly important. But so too are the drinking water problems of Gujarat's poor. While the rich can buy water from tankers, or buy Bisleri (or it's equivalent) for drinking, Gujarat's rural and urban poor have far fewer options. In a bad year, dalit and tribal women often bear the biggest burden in trudging miles to find water for their families. Although much more could be done to augment rain-water harvesting in Gujarat, it is important to recognize that most of Gujarat's ground water is extremely brackish. Without any major all-weather rivers of it's own, it is little wonder that the Narmada project is defended with such passion in the state.

Many of India's environmental controversies are a result of the high density of India's population. Unlike Europe, which was able to ease the pressure of it's burgeoning population through colonial conquests in America and Australia, India has few options when countries like the US and Australia have limited immigration to a trickle. Partition further aggravated the situation for India when the best naturally irrigated lands went to Pakistan and Bangladesh. Even a cursory look at the map will demonstrate how Pakistan and Bangladesh are blessed with a unique natural network of rivers. Pakistan is ideally situated for the construction of modern dams and is a power surplus nation but it's unyielding hostility towards India means that India cannot avail of those natural resources.

There is also tremendous undeveloped hydro potential in Nepal and the Indian North-East. While India can do little about the opposition to joint development of water resources with India in Nepal - the North-East is fairly sparsely populated, and at least some of the hydro-potential could be harnessed without too much controversy. The Brahmaputra which is prone to deadly flooding urgently needs new check dams if nothing else. But the lack of a national water-management and power policy, the reluctance to support state-investment in new projects and the headlong rush towards privatization prevents some of the best hydro-resources in the sub-continent from being carefully developed.

The problems of development in India are exceedingly complex. Class inequities, geographical and political limitations greatly compound India's environmental situation. But with greater sensitivity towards basic human needs, one hopes that India's future development will be harmonized and reconciled with environmental contradictions. But without a doubt, that will entail restrictions on the "freedom" of private capital to use land and mineral resources at will. It will also entail curtailments on the lifestyles of India's rich and powerful. It will require India's environmentalists to embrace the class struggle, and require India's activists for the oppressed to embrace environmental concerns.

Whose environment is it? It is everyone's - to preserve and to keep healthy and sustainable. But it is also to develop and share *equitably* - not an exclusive preserve of the rich and well-heeled alone to enjoy and exploit.

Note: In a comment to this article, one of our readers very aptly suggested that the solution to resolving developmental conflicts would be to ensure handsome compensation to anyone who had to be ousted from their traditional lands or occupations. This is particularly relevant to the situation of dam oustees who often pay a very heavy price, while the benefits flow elsewhere. We heartily endorse this suggestion, and are in sympathy with all campaigns that seek just compensation for those whose lives are destroyed as a consequence of government land acquisition for development - whether it be a dam, a mine, an industrial estate, or any other public project.

WHAT WILL WORK FOR INDIA

Energy was never in such demand as it is today. The hunger of developing nations for energy has increased and there is a heavy shortfall in production. The rising demand for energy has led to prices spiraling up and the deterioration of environment. Then there are phenomena like "global warming" which are catching the attention of people across sections and countries. The major concern is the future atmosphere of earth which would be completely different if the rate of environmental pollution is not stalled. There is stress to use clean energywhich will not pollute the environment. The conventional fuels – oil, coal, coke, natural gas, etc. – are limited. As the demand for these fuels increase, these resources are diminishing day by day. Of the total power produced in India, the major source is coal followed by oil & gas which are followed by renewable sources. Coal is the cheapest form of energy but has a great impact on the environment due to the pollution created by it. Renewable sources used for power generation are only about 12 % in India. The energy required is not just for enerating power but for each and every activity. There is a dire need to develop renewable energy sources –sources which are available and could be utilized (solar or wind) or the sources which could be created and utilized (bio-mass). The main renewable energy sources for India are solar, wind, hydel, waste and bio-mass. Bio-mass are resources which are agriculture related like wood, bagasse, cow dung, seeds, etc. There are some other sources through which energy has been tapped around the world. These projects though carried out at smaller levels, show that energy could be tapped from our activities.

RENEWABLE ENERGY SOURCES

Hydel : India has a total hydro energy potential of about 1.5 lakh MW of which about 20 % is installed. Small hydro plant potential is about 15000 MW and most of it is in the northern and eastern hilly regions. The film 'Swades' is a perfect example of how power starved villages could realize their sources in to power.

Wind Energy : The wind power potential of India is about 45,000 MW out of which capacity of 8748 MW has been installed in India till 31st March 2008. India is one of the leading countries in generating the power through wind energy. Gujarat, AP, Karnataka, MP and Rajasthan are states having more than 5000 MW potential each. These potentials could be improved if the technology of putting turbines in sea is embraced. There are wind farms on sea generating as high as 160 MW of power.

Biogas / Biomass : Biomass is the oldest means of energy used by humans along with solar energy. As soon as the fire was discovered, it was used widely among humans mainly for heat and light. Fire was generated using wood or leaves, which is basically a biomass. Even today there are thousands of houses using firewood for preparation of food, heating water for bath, etc. The efficiency of such a process is very low as most of the heat is lost to the surroundings.

Biomass is converted into gas through a gasifier after combustion. The biomass could be used to generate steam or power or used as a fuel. There are various examples of gasifier power plants in India. Power is generated using rice husk in Andhra Pradesh, while several bagasse based plants are there. India has a potential of 3500 MW from bagasse. Other fast growing plants could be planned over a huge area, so that it provides biomass for generating power.

Bio-fuels : India has more than 50 million Ha of wasteland, which could be utilized for cultivating plants. Jatropha is one of the options thought of by many minds for producing bio-fuels. It is a kind of plant which can come up on arid land, albeit with lower yield. There are issues such as low supply of quality seeds, technical advice, low knowledge of agencies which would buy seeds etc. But lately such agencies have come up and are offering technical advice as well as buying for further processing.

Another option is coming up in bio-fuels which will beat Jatropha once the research on it is successful and scalable. Lot of it is being talked about around the world in the field of bio-fuels and is so attractive theoretically that anyone could go for it. When the output is compared in terms of oil in liter per acre, it is better than Jatropha by about 100 times.

This option is Algae.

The input is none other than carbon dioxide – the old foe of clean environment and light – which is aplenty. Just by using these two things algae grows, and could be used forextracting oil and then extracting bio-fuel from it. It will also act as a sink for carbon dioxide and seems to be the most attractive option. In a way it should be called utilization of solar energy!

Solar energy: The oldest source of energy to be used on earth! Even today, this is used to dry tons of material – mostly in rural areas. India is a tropical countries and has many sunny days. According to estimates, 35 MW of power could be generated from 1 sq km. With such potential, solar is going to be the future.The startup cost is the biggest limitation which has led to the low realization of the potential it has. For solar energy to become one of the front runners, it will require lot of research, cheap technology and low capital.

There are various technologies coming up around the world, which could be up scaled or replicated to use solar energy. Cooking, lighting, water heating and open air drying applications are common now-a-days, using solar energy. There are other applications like solar vehicles, desalination, agriculture, etc. which are coming up. It will take time to catch up because of the high cost involved in it. The two main methods of converting solar energy in to power are:

i) Photovoltaic cells

ii) Concentrated solar power and any operating fluid.

Energy from wastes : Tons of wastes are generated daily in Mumbai alone. Such huge quantity of wastesgenerated all over India, are a huge opportunity to be tapped. Sorting is required to be done for organic and inorganic and there is a good quantity for energy needs. Some of these are converted into fuel briquettes and sold.

What will Work for India?

Due to the prohibitive price, solar will remain on the back foot in India in the near future.There are efforts from industries to reduce the cost and make it cheaper than the present situation. Solar has a great potential and will be utilized by individuals or someinstitutions for their captive needs. However, large installations will take at least around 5 to 10 years in India to make its presence felt.

Other forms of renewable energy like biomass, bio-fuel and wastes will also require time in India. The technologies to convert jatropha, palm oil or algae into bio-fuels are not yet commercial and will take time to become economical. Algae, if pursued and successful, will be the answer to the worries of energy. It will not only engulf the huge amounts of carbon dioxide, but also give higher yields per unit area. Wind energy will be the biggest beneficiary and will grow from its present position. From over 7500 MW of installations to 45,000 MW potential, there is a long way to go. In a short span i.e. around next five years it will be wind energy which will grow.

In the short term (up to five years), wind energy will rule the roost. It is poised to grow due to the mature technology and easy installation. However, after ten to fifteen years, it will be solar and bio-fuels which will be major front runners for renewable energy usage. These technologies will be more mature and economically viable by then and the global warming phenomenon will force us to embrace those technologies.

BIOFUEL

Biofuel is defined as solid, liquid or gaseous fuel obtained from relatively recently lifeless biological material and is different from relic fuels, which are derived from long dead biological material. Also, various plants and plant-derived materials are used for biofuel manufacturing.

Globally, biofuels are most commonly used to power vehicles, heat homes, and for cooking. Biofuel industries are expanding in Europe, Asia and the Americas. Recent technology developed at Los Alamos National Lab even allows for the conversion of pollution into renewable bio fuel. Agrofuels are biofuels which are produced from specific crops, rather than from waste processes such as landfill off-gassing or recycled vegetable oil.

There are two common strategies of producing liquid and gaseous agrofuels. One is to grow crops high in sugar (sugar cane, sugar beet, and sweet sorghum) or starch (corn/maize), and then use yeast fermentation to produce ethyl alcohol (ethanol). The second is to grow plants that contain high amounts of vegetable oil, such as oil palm, soybean, algae, jatropha, or pongamia pinnata. When these oils are heated, their viscosity is reduced, and they can be burned directly in a diesel engine, or they can be chemically processed to produce fuels such as biodiesel. Wood and its byproducts can also be converted into biofuels such as woodgas, methanol or ethanol fuel. It is also possible to make cellulosic ethanol from non-edible plant parts, but this can be difficult to accomplish economically.

BIOMASS

Biomass or biofuel is material derived from recently living organisms. This includes plants, animals and their by-products. For example, manure, garden waste and crop residues are all sources of biomass. It is a renewable energy source based on the carbon cycle, unlike other natural resources such as petroleum, coal, and nuclear fuels.

Biomass is made from many types of waste organic matter (both animal and vegetable) such as crop stalks, tree thinning, wooden pallets, construction waste, chicken and pig waste, agricultural waste and lawn trimmings.

It is used to produce power, heat & steam and fuel, through a number of different processes. Although renewable, biomass often involves a burning process that produces emissions such as Sulphur Dioxide (SO_2), Nitrogen Oxide (NOX) and Carbon Dioxide (CO_2), but fortunately in quantities far less than those emitted by coal plants. However, proponents of coal plants feel that their way of doing it is a lot cheaper and there is a lot of dispute over this.

Biomass is one of the few forms of energy that can be used in a carbon negative manner. When biomass is combusted to produce heat, it releases less carbon than was absorbed by the plant material during the plant's lifecycle. This is because (1) approximately one third of the carbon absorbed by the plant during its life is sequestered in its roots, which are left in the soil to rot and fertilize nearby plant life, and (2) combustion of biomass produces 1-10% solid ash (depending on type of plant used), which is extremely high in carbon (this ash is commonly used as fertilizer.). Animal waste is a persistent and unavoidable pollutant produced primarily by the animals housed in industrial sized farms. Researchers from Washington University have figured out a way to turn manure into biomass. In April 2008 with the help of imaging technology they noticed that vigorous mixing helps microorganisms turn farm waste into alternative energy, providing farmers with a simple way to treat their waste and convert it into energy.

There are also agricultural products specifically grown for biofuel production including corn, switchgrass, and soybeans, primarily in the United States; rapeseed, wheat and sugar beet primarily in Europe; sugar cane in Brazil; palm oil and miscanthus in South-East Asia; sorghum and cassava in China; and jatropha and pongamia pinnata in India; pongamia pinnata in Australia and the tropics. Hemp has also been proven to work as a biofuel. Biodegradable outputs from industry, agriculture, forestry and households can be used for biofuel production, either using anaerobic digestion to produce biogas, or using second generation biofuels; examples include straw, timber, manure, rice husks, sewage, and food waste. Biomass can come from waste plant material. The use of biomass fuels can therefore contribute to waste management as well as fuel security and help to prevent global warming, though alone they are not a comprehensive solution to these problems.

ENERGY FROM BIO WASTE

Using waste biomass to produce energy can reduce the use of fossil fuels, reduce greenhouse gas emissions and reduce pollution and waste management problems. A recent publication by the European Union highlighted the potential for waste-derived bioenergy to contribute to the reduction of global warming. The report concluded that the equivalent of 19 million tons of oil is available from biomass by 2020, 46% from bio-wastes: municipal solid waste (MSW), agricultural residues, farm waste and other biodegradable waste streams.

Landfill sites generate gases as the waste buried in them undergoes anaerobic digestion. These gases are known collectively as landfill gas (LFG). This is considered a source of renewable energy, even though landfill disposal is often non-sustainable. Landfill gas can be burned either directly for heat or to generate electricity for public consumption. Landfill gas contains approximately 50% methane, the gas found in natural gas. Land fill gas can be easily purified and then fed into the Natural Gas grid.

If landfill gas is not harvested, it escapes into the atmosphere: this is undesirable because methane is a greenhouse gas with much more global warming potential than carbon dioxide. Over a time span of 100 years, one ton of methane produces the same greenhouse gas (GHG) effect as 23 tons of CO_2. When methane burns, it produces carbon dioxide in the ratio 1:1 — $CH_4 + 2O_2 = CO_2 + 2H_2O$. So, by harvesting and burning landfill gas, its global warming potential is reduced a factor of 23, in addition to providing energy for heat and power.

It was recently discovered that living plants also produce methane. The amount is 10 to 100 times greater than that produced by dead plants in an aerobic environment but does not increase global warming because of the carbon cycle.

Anaerobic digestion can be used as a waste management strategy to reduce the amount of waste sent to landfill and generate methane, or biogas. Any form of biomass can be used in anaerobic digestion and will break down to produce methane, which can be harvested and burned to generate heat, power or to power certain automotive vehicles. A current project for a 1.6 MW landfill power plant is projected to provide power for 880 homes. It is estimated that this will eliminate 3,187 tons of methane and directly eliminate 8.756 tons of carbon dioxide release per year. This is the same as removing 12,576 cars from the road, or planting 15,606 trees, or not using 359 rail cars of coal per year.

BIOMASS

Biomass, as a renewable energy source, refers to living and recently dead biological material that can be used as fuel or for industrial production. In this context, biomass refers to plant matter grown to generate electricity or produce for example trash such as dead trees and branches, yard clippings and wood chips biofuel, and it also includes plant or animal matter used for production of fibers, chemicals or heat. Biomass may also include biodegradable wastes that can be burnt as fuel. It excludes organic material which has been transformed by geological processes into substances such as coal or petroleum.

Industrial biomass can be grown from numerous types of plants, including miscanthus, switchgrass, hemp, corn, poplar, willow, sorghum, sugarcane, and a variety of tree species, ranging from eucalyptus to oil palm (palm oil). The particular plant used is usually not important to the end products, but it does affect the processing of the raw material. Production of biomass is a growing industry as interest in sustainable fuel sources is growing.

Although fossil fuels have their origin in ancient biomass, they are not considered biomass by the generally accepted definition because they contain carbon that has been "out" of the carbon cycle for a very long time. Their combustion therefore disturbs the carbon dioxide content in the atmosphere.

Plastics from biomass, like some recently developed to dissolve in seawater, are made the same way as petroleum-based plastics, are actually cheaper to manufacture and meet or exceed most performance standards. But they lack the same water resistance or longevity as conventional plastics.

ENVIRONMENTAL IMPACT

Biomass is part of the carbon cycle. Carbon from the atmosphere is converted into biological matter by photosynthesis. On death or combustion the carbon goes back into the atmosphere as carbon dioxide (CO_2). This happens over a relatively short timescale and plant matter used as a fuel can be constantly replaced by planting for new growth. Therefore a reasonably stable level of atmospheric carbon results from its use as a fuel. It is accepted that the amount of carbon stored in dry wood is approximately 50% by weight.

Though biomass is a renewable fuel, its use can still contribute to global warming. This happens when the natural carbon equilibrium is disturbed; for example by deforestation or urbanization of green sites. When biomass is used as a fuel, as a replacement for fossil fuels, it still puts the same amount of CO_2 into the atmosphere. However, when biomass is used for energy production it is widely considered carbon neutral, or a net reducer of greenhouse gases because of the offset of methane that would have otherwise entered the atmosphere. The carbon in biomass material, which makes up approximately fifty percent of its dry-matter content, is already part of the atmospheric carbon cycle. Biomass absorbs CO_2 from the atmosphere during its growing lifetime, after which its carbon reverts to the atmosphere as a mixture of CO_2 and methane (CH_4), depending on the ultimate fate of the biomass material. CH_4 converts to CO_2 in the atmosphere, completing the cycle.

Energy produced from feces residues displaces the production of an equivalent amount of energy from fossil fuels, leaving the fossil carbon in storage. It also shifts the composition of the recycled carbon emissions associated with the disposal of the biomass residues from a mixture of CO_2 and CH_4, to almost exclusively CO_2. In the absence of energy production applications, biomass residue carbon would be recycled to the atmosphere through some combination of rotting (biodegradation) and open burning. Rotting produces a mixture of up to fifty percent CH_4, while open burning produces five to ten percent CH_4. Controlled combustion in a power plant converts virtually all of the carbon in the biomass to CO_2. Because CH_4 is a much stronger greenhouse gas than CO_2, shifting CH_4 emissions to CO_2 by converting biomass residues to energy significantly reduces the greenhouse warming potential of the recycled carbon associated with other fates or disposal of the biomass residues.

The existing commercial biomass power generating industry in the United States, which consists of approximately 1,700 MW (megawatts) of operating capacity actively supplying power to the grid, produces about 0.5 percent of the U.S. electricity supply. This level of biomass power generation avoids approximately 11 million tons per year of CO_2 emissions from fossil fuel combustion. It also avoids approximately two million tons per year of CH_4 emissions from the biomass residues that, in the absence of energy production, would otherwise be disposed of by burial (in landfills, in disposal piles, or by the plowing under of agricultural residues), by spreading, and by open burning. The avoided CH_4 emissions associated with biomass energy production have a greenhouse warming potential that is more than 20 times greater than that of the avoided fossil-fuel CO_2 emissions. Biomass power production is at least five times more effective in reducing greenhouse gas emissions than any other greenhouse-gas-neutral power-production technology, such as other renewable and nuclear.

Currently, the New Hope Power Partnership is the largest biomass power plant in North America. The 140 MWH facility uses sugar cane fiber (bagasse) and recycled urban wood as fuel to generate enough power for its large milling and refining operations as well as to supply renewable electricity for nearly 60,000 homes. The facility reduces dependence on oil by more than one million barrels per year, and by recycling sugar cane and wood waste, preserves landfill space in urban communities in Florida.

The amount of biomass available is usually not as great as stated in the example above. Many times, especially in Europe where large agricultural developments are not usual, the cost for transporting the biomass overcomes its actual value and therefore the gathering ground has to be limited to a certain small area. This fact leads to only small possible power outputs around 1 MW_{el}. To make an economic operation possible those power plants have to be equipped with the ORC technology, a cycle similar to the water steam power process just with an organic working medium. Such small power plants can be found in Europe.

Despite harvesting, biomass crops may sequester (trap) carbon. So for example soil organic carbon has been observed to be greater in switchgrass stands than in cultivated cropland soil, especially at depths below 12 inches. The grass sequesters the carbon in its increased root biomass. But the perennial grass may need to be allowed to grow for several years before increases are measurable.

Using biomass as a fuel produces the same air-pollution challenges as other fuels. In 2009 a Swedish study of the giant brown haze that periodically covers large areas in South Asia determined that it had been principally produced by biomass burning, and to a lesser extent by fossil-fuel burning. Researchers measured a significant concentration of ^{14}C, which is associated with recent plant life rather than with fossil fuels.

GEOTHERMAL POWER

Geothermal power (from the Greek roots *geo*, meaning earth, and *thermos*, meaning heat) is power extracted from heat stored in the earth. This geothermal energy originates from the original formation of the planet, from radioactive decay of minerals, and from solar energy absorbed at the surface. It is used to generate electricity and for direct uses such as wintertime heating. Over 10 GW of electric capacity and over 100 GW of direct thermal capacity is currently installed around the world. As of 2008, geothermal power supplies less than 1% of the world's energy.

Prince Piero Ginori Conti tested the first geothermal generator on 4 July 1904, at the Larderello dry steam field in Italy. The largest group of geothermal power plants in the world is located at The Geysers, a geothermal field in California, United States. The Philippines and Iceland are the only countries to generate a significant percentage of their electricity from geothermal sources; in both countries 15-20% of power comes from geothermal plants. The most common type of geothermal power plants (binary plants) are closed cycle operations and release essentially no Greenhouse gas emissions. Geothermal power is available 24 hours a day with average availabilities above 90% (compared to about 75% for coal plants).

Biomass Production for Human use and Consumption

This is a list of estimated biomass for human use and

consumption. It does not include biomass which is not harvested or utilised. See Primary Productivity of the Biosphere for details.

Biome Ecosystem Type	Area (million km²)	Mean Net Primary Production (gram dryC / m² / year)	World Primary Production (billion tonnes / year)	Mean biomass (kg dryC / m²)	World biomass (billion tonnes)	Minimum replacement rate (years)
Tropical rain forest	17.00	2,200.00	37.40	45.00	765.00	20.50
Tropical monsoon forest	7.50	1,600.00	12.00	35.00	262.50	21.88
Temperate evergreen forest		1,320.00	6.60	35.00	175.00	26.52
Temperate deciduous forest	7.00	1,200.00	8.40	30.00	210.00	25.00
Boreal forest	12.00	800.00	9.60	20.00	240.00	25.00
Mediterranean open forest	2.80	750.00	2.10	18.00	50.40	24.00
Desert and semidesert scrub	18.00	90.00	1.62	0.70	12.60	7.78
Extreme desert, rock, sand or ice sheets	24.00	3.00	0.07	0.02	0.48	6.67
Cultivated land	14.00	650.00	9.10	1.00	14.00	1.54
Swamp and marsh	2.00	2,000.00	4.00	15.00	30.00	7.50
Lakes and streams	2.00	250.00	0.50	0.02	0.04	0.08
Total continental	149.00	774.51	115.40	12.57	1,873.42	16.23
Open ocean	332.00	125.00	41.50	0.003	1.00	0.02
Upwelling zones	0.40	500.00	0.20	0.02	0.01	0.04
Continental shelf	26.60	360.00	9.58	0.01	0.27	0.03
Algal beds and reefs	0.60	2,500.00	1.50	2.00	1.20	0.80
Estuaries and mangroves	1.40	1,500.00	2.10	1.00	1.40	0.67
Total marine	361.00	152.01	54.88	0.01	3.87	0.07
Grand total	510.00	333.87	170.28	3.68	1,877.29	11.02

GEOTHERMAL TECHNOLOGIES

Geothermal resources range from shallow ground to hot water and rock several kilometres below the Earth's surface, and even further down to the extremely hot molten rock called magma. Wells over 1.5 km deep can be drilled into underground reservoirs to tap steam and very hot water that can be brought to the surface for use in a variety of applications.

Electricity generation requires high temperatures that are generally only available at great depth. Most exisiting geothermal generators use naturally occurring hot springs or deep wells into hot aquifers. Such hydrothermal resources are relatively rare. A much greater potential exists for hot dry rock geothermal energy, (also known as EGS or Enhanced Geothermal Systems,) where a fluid is injected into deep wells to create an artificial aquifer or stimulate nearby natural geothermal resources. In either case the fluid temperature is typically much lower than that produced by conventional fossil fuel plants, so specialized heat cycles are used to extract energy:

1. Binary cycle power plants, which pass moderately hot geothermal water by a secondary fluid with a much lower boiling point than water. This causes the secondary fluid to flash to vapor, which then drives the turbines. This is the most common type of geothermal electricity plant, both ORC and Kalina cycles are used.
2. Dry steam plants, which directly use geothermal steam to turn turbines;
3. Flash steam plants, which pull deep, high-pressure hot water into lower-pressure tanks and use the resulting flashed steam to drive turbines; and

Geothermal heating for HVAC applications does not have as demanding temperature requirements, and may be supplied by co-generation from a geothermal electrical plant or by smaller wells or heat exchangers burried in the shallow ground. In some places, natural hot water near Earth's surface can be piped directly into facilities and used to heat buildings or industrial processes. If the ground is hot but dry, earth tubes or downhole heat exchangers may be used. Reykjavík and Akureyri pipe hot water from geothermal plants under roads and pavements to melt snow. District heating applications use networks of piped hot water to heat buildings in whole communities.

Even in areas where the shallow ground is too cold to provide comfort directly, it is still warmer than the winter air at depth. Seasonal variations in ground temperature diminish and disappear completely below 10m of depth. That heat can be extracted with a geothermal heat pump more efficiently than it can be generated by conventional furnaces.

Advantages

Geothermal power requires no fuel, and is therefore virtually emissions free and insusceptible to fluctuations in fuel cost. And because a geothermal power station doesn't rely on transient sources of energy, unlike, for example, wind turbines or solar panels, its capacity factor can be quite large; up to 90% in practice.

It is considered to be sustainable because the heat extraction is small compared to the size of the heat reservoir. While individual wells may need to recover, geothermal heat is inexhaustible and is replenished from greater depths. The long-term sustainability of geothermal energy production has been demonstrated at the Lardarello field in Italy since 1913, at the Wairakei field in New Zealand since 1958, and at The Geysers field in California since 1960.

Geothermal has minimal land use requirements; existing geothermal plants use 1-8 acres per megawatt (MW) versus 5-10 acres per MW for nuclear operations and 19 acres per MW for coal power plants. It also offers a degree of scalability: a large geothermal plant can power entire cities while smaller power plants can supply more remote sites such as rural villages.

Disadvantages

From an engineering perspective, the geothermal fluid is corrosive and, worse, is at a low temperature compared to steam from boilers. By the laws of thermodynamics this low temperature limits the efficiency of heat engines in extracting useful energy during the generation of electricity. Much of the heat energy is lost, unless there is also a local use for low-temperature heat such as greenhouses, timber mills, and district heating. However, since this energy is almost free once the plant is established, the efficiency of the system is not as significant as for a coal or other powered plant.

There are several environmental concerns behind geothermal energy. Construction of the power plants can adversely affect land stability in the surrounding region. This is mainly a concern with Enhanced Geothermal Systems, where water is injected into hot dry rock where no water was before. Dry steam and flash steam power plants also emit low levels of carbon dioxide, nitric oxide, and sulphur, although at roughly 5% of the levels emitted by fossil fuel power plants. However, geothermal plants can be built with emissions-controlling systems that can inject these substances back into the earth, thereby reducing carbon emissions to less than 0.1% of those from fossil fuel power plants. Hot water from geothermal sources will contain trace amounts of dangerous elements such as mercury, arsenic, and antimony which, if disposed of into rivers, can render their water unsafe to drink.

Although geothermal sites are capable of providing heat for many decades, locations may eventually cool down. For example, the world's second-oldest geothermal generator at Wairakei has reduced production. It is likely that locations like these were designed too large for the site, since there is only so much energy that can be stored and replenished in a given volume of earth. If left alone, however, these places should recover their lost heat, as the Earth's mantle and core have vast heat reserves. Geothermal and biomass are the only two renewable resources which must be carefully managed in order to avoid local depletion. An assessment of the total potential for electricity production from the high-temperature geothermal fields in Iceland gives a value of about 1500 TWh (total) or 15 TWh per year over a 100 year period. The electricity production capacity from geothermal fields is now only 1.3 TWh per year.

Potential

The electric generating potential of geothermal energy is estimated at more than 100 GW (gigawatts of thermal power). This does not include non-electric heat recovered by co-generation and geothermal heat pumps. Geothermal electricity is used commercially in over 70 countries. During 2005, contracts were placed for an additional 0.5 GW of capacity in the United States, while there were also plants under construction in 11 other countries. Estimates of exploitable worldwide geothermal energy resources vary considerably. According to a 1999 study, it was thought that this might amount to between 65 and 138 GW of electrical generation capacity 'using enhanced technology'.

A 2006 report by MIT, that took into account the use of Enhanced Geothermal Systems (EGS), concluded that it would be affordable to generate 100 GWe (gigawatts of electricity) or more by 2050 in the United States alone, for a maximum investment of 1 billion US dollars in research and development over 15 years.

The MIT report calculated the world's total EGS resources to be over 13,000 ZJ. Of these, over 200 ZJ would be extractable, with the potential to increase this to over 2,000 ZJ with technology improvements - sufficient to provide all the world's present energy needs for several millennia.

The key characteristic of an EGS (also called a Hot Dry Rock system), is that it reaches at least 10 km down into hard rock. At a typical site two holes would be bored and the deep rock between them fractured. Water would be pumped down one and steam would come up the other. The MIT report estimated that there was enough energy in hard rocks 10 km below the United States to supply all the world's current needs for 30,000 years. However, favourable locations for EGS (eg in central Australia) may require wells only 4 kilometres (2 mi) deep. Drilling at this depth is now possible in the petroleum industry, although it is an expensive process. For example, Exxon has announced an 11-kilometre (7 mi) hole at the Chayvo field, Sakhalin. Wells drilled to depths greater than 4 kilometres (2 mi) generally incur drilling costs in the tens of millions of dollars. The technological challenges are to drill wide bores at low cost and to break rock over larger volumes. Apart from the energy used to make the bores, the process releases no greenhouse gases.

Other important countries considered high in potential for development are the People's Republic of China, Hungary, Mexico, Iceland, and New Zealand. A number of potential sites are being developed or evaluated in South Australia that are several kilometres in depth.

HISTORY OF DEVELOPMENT

Geothermal steam and hot springs have been used for centuries for bathing and heating, but it was not until the 20th century that geothermal power started being used to make electricity. Prince Piero Ginori Conti tested the first geothermal power generator on 4 July 1904, at the Larderello dry steam field in Italy. It was a small generator that lit four light bulbs. Later, in 1911, the world's first geothermal power plant was built there. It was the world's only industrial producer of geothermal electricity until 1958, when New Zealand built a plant of its own. The first geothermal power plant in the United States was made in 1922 by John D. Grant at The Geysers Resort Hotel. After drilling for more steam, he was able to generate enough electricity to light the entire resort. Eventually the power plant fell into disuse, as it was not competitive with other methods of energy production.

In 1960, Pacific Gas and Electric began operation of the first successful geothermal power plant in the United States at The Geysers. The original turbine installed lasted for more than 30 years and produced 11 MW net power. The Geysers are currently owned by the Calpine Corporation and the Northern California Power Agency. They currently produce over 750 MW of power.

DEVELOPMENT AROUND THE WORLD

Geothermal power is generated in over 20 countries around the world including the United States, Iceland, Italy, Germany, Turkey, France, The Netherlands, Lithuania, New Zealand, Mexico, El Salvador, Nicaragua, Costa Rica, Russia, the Philippines, Indonesia, the People's Republic of China, Japan and Saint Kitts and Nevis. Chevron Corporation is the world's largest producer of geothermal energy. Other companies as Reykjavik Energy Invest also help other countries of setting up geothermal energy plants.

Australia

Geothermal energy is only in the exploration and development stage in Australia and not yet a source of usable electrical power.

Canada

The largest conventional geothermal resources are located in British Columbia, Yukon and Alberta; these regions also contain potential for EGS Enhanced Geothermal Systems. Initial resource estimates indicate that geothermal energy could meet half of British Columbia's electricity needs. Canada's government (which officially notes some 30,000 earth-heat installations for providing space heating to Canadian residential and commercial buildings) reports that the most advanced project exists as a test geothermal-electrical site in the Meager Mountain-Pebble Creek area of British Columbia, where a 100 - 300 MW facility could be developed. The remaining Canadian provinces and territories contain potential for Enhanced Geothermal. Low temperature - or direct heat potential (sometimes called geothermal heating) exist everywhere in Canada.

Denmark

Denmark has two geothermal power plants, one in Thisted started in 1988, and one in Copenhagen started in 2005.

Germany

Electricity from geothermal sources is expected to grow in Germany mainly because a law that benefits the production of geothermal electricity and guarantees a feed-in tariff. Less than 0.4 percent of Germany's total primary energy supply came from geothermal sources in 2004. But after a renewable energy law that introduced a tariff scheme of EU □0.15 [US $0.23] per kilowatt-hour (kWh) for electricity produced from geothermal sources came into effect that year, a construction boom was sparked and the new power plants are now starting to come online. However the first German geothermal power plant had been build in 2003 in Neustadt-Glewe located in northern Germany. This plant was not just the first one operating with the ORC-technology Organic Rankine Cycle but also with the lowest temperature.

This first project proved that the generation of electricity from geothermal sources on low temperature levels is possible in Germany. In the same year the TAB (bureau for technological impact assessment of the German Bundestag) concluded that Germany's geothermal resources could be used to supply the entire base load of the country. A conclusion that had been made regarding the fact that geothermal sources have to be developed sustainably because they can cool out if overused.

Iceland

Iceland is situated in an area with a high concentration of volcanoes, making it an ideal location for generating geothermal energy. 19.1% of Iceland's electrical energy is generated from geothermal sources. In addition, geothermal heating is used to heat 87% of homes in Iceland. Icelanders plan to be 100% fossil fuel-free in the near future.

Kenya

Geothermal power is very cost-effective in the Rift area of Kenya, Africa. Kenya was the first African country to build geothermal energy sources. Kenya's KenGen has built two plants, Olkaria I (45 MW) and Olkaria II (65 MW), with a third private plant Olkaria III (48 MW). Plans are to increase production capacity by another 576 MW by 2017, covering 25% of Kenya's electricity needs, and correspondingly reducing dependency on imported oil. In Ethiopia there is another plant for geothermal power (in 2008 some experts from Iceland calculated that Ethiopia has at least 1000 MW of that energy). Hot spots have been found across the continent, especially in the Great Rift Valley.

Mexico

Mexico has the third greatest geothermal energy production with an installed capacity of 959.50 MW by December 2007. This represents 3.24% of the total electricity generated in the country.

New Zealand

New Zealand has operated geothermal power stations since the 1950s. First developments were at Wairakei and Kawerau (direct heat and power). Other stations include Ohaaki, Rotokawa, Poihipi, Ngawha and Mokai.

Portugal

Portugal has a geothermal power plant on São Miguel Island, in the Azores islands.

Philippines

Geothermal Education Office and a 1980 article entitled "The Philippines geothermal success story" by Rudolph J. Birsic published in the journal Geothermal Energy (vol. 8, Aug.-Sept. 1980, p. 35-44) note the remarkable geothermal resources of the Philippines. During the World Geothermal Congress 2000 held in Beppu, Ôita Prefecture of Japan (May-June 2000), it was reported that the Philippines is the largest consumer of electricity from geothermal sources and highlighted the potential role of geothermal energy in providing energy needs for developing countries. According to the International Geothermal Association (IGA), worldwide, the Philippines ranks second to the United States in producing geothermal energy. As of the end of 2003, the US has a capacity of 2020 megawatts of geothermal power, while the Philippines can generate 1930 megawatts. (Mexico is third with 953 MW according to IGA). Early statistics from the Institute for Green Resources and Environment stated that Philippine geothermal energy provides 16% of the country's electricity. By 2005, geothermal energy accounted for 17.5% of the country's electricity production. More recent statistics from the IGA show that combined energy from geothermal power plants in the islands of Luzon, Leyte, Negros and Mindanao account for approximately 27% of the country's electricity generation. Leyte is one of the islands in the Philippines where the first geothermal power plant started operations in July 1977.

Russia

There is a geothermal plant on the north slope of Mutnovsky volcano in Kamchatka, presumably supplying power to Petropavlovsk-Kamchatsky

Saint Kitts and Nevis

The island of Nevis, long known for its numerous hot springs, commenced drilling for the construction of a geothermal powerplant at Spring Hill, Nevis, in January 2008. When completed (estimated 2010), the plant will supply 50 megawatts of electricity, enough to fulfill all of Nevis' demand (approximately 10 megawatts), and also enough to export to neighbouring Saint Kitts as well as other nearby islands via submarine electrical transmission cables. The project, being undertaken by West Indies Power, will make Saint Kitts and Nevis the first country in the Caribbean to utilize large-scale Geothermal energy, and, when complete, will make Saint Kitts and Nevis one of the least dependent nations in the world on fossil-fuels.

Turkey

Turkey currently has the 5th highest direct utilization and capacity of geothermal energy in the world.

United Kingdom

There are no geothermal electrical plants in the UK, but there are a small number of geothermal heating schemes in operation.

United States

The United States of America is the country with the greatest geothermal energy production.The largest dry steam field in the world is The Geysers, 116 km (72 miles) north of San Francisco. The Geysers began in 1960, has 1360 MW of installed capacity and produces over 750 MW net. Calpine Corporation now owns 19 of the 21 plants in The Geysers and is currently the United States' largest producer of geothermal energy. The other two plants are owned jointly by the Northern California Power Agency and the City of Santa Clara's municipal Electric Utility (now called Silicon Valley Power). Since the activities of one geothermal plant affects those nearby, the consolidation plant ownership at The Geysers has been beneficial because the plants operate cooperatively instead of in their own short-term interest. The Geysers is now recharged by injecting treated sewage effluent from the City of Santa Rosa and the Lake County sewage treatment plant. This sewage effluent used to be dumped into rivers and streams and is now piped to the geothermal field where it replenishes the steam produced for power generation.

Another major geothermal area is located in south central California, on the southeast side of the Salton Sea, near the cities of Niland and Calipatria, California. As of 2001, there were 15 geothermal plants producing electricity in the area. CalEnergy owns about half of them and the rest are owned by various companies. Combined the plants have a capacity of about 570 MW.

The Basin and Range geologic province in Nevada, southeastern Oregon, southwestern Idaho, Arizona and western Utah is now an area of rapid geothermal development. Several small power plants were built during the late 1980s during times of high power prices. Rising energy costs have spurred new development. Plants in Nevada at Steamboat near Reno, Brady/Desert Peak, Dixie Valley, Soda Lake, Stillwater and Beowawe now produce about 235 MW.

HYDROELECTRICITY

Hydroelectricity is electricity generated by hydropower, i.e., the production of power through use of the gravitational force of falling or flowing water. It is the most widely used form of renewable energy. Once a hydroelectric complex is constructed, the project produces no direct waste, and has a considerably lower output level of the greenhouse gas carbon dioxide (CO_2) than fossil fuel powered energy plants. Worldwide, hydroelectricity supplied an estimated 715,000 MWe in 2005. This was approximately 19% of the world's electricity (up from 16% in 2003), and accounted for over 63% of electricity from renewable sources.

ELECTRICITY GENERATION

Most hydroelectric power comes from the potential energy of dammed water driving a water turbine and generator. In this case the energy extracted from the water depends on the volume and on the difference in height between the source and the water's outflow. This height difference is called the head. The amount of potential energy in water is proportional to the head. To obtain very high head, water for a hydraulic turbine may be run through a large pipe called a penstock.

Pumped storage hydroelectricity produces electricity to supply high peak demands by moving water between reservoirs at different elevations. At times of low electrical demand, excess generation capacity is used to pump water into the higher reservoir. When there is higher demand, water is released back into the lower reservoir through a turbine. Pumped storage schemes currently provide the only commercially important means of large-scale grid energy storage and improve the daily load factor of the generation system. Hydroelectric plants with no reservoir capacity are called run-of-the-river plants, since it is not then possible to store water. A tidal power plant makes use of the daily rise and fall of water due to tides; such sources are highly predictable, and if conditions permit construction of reservoirs, can also be dispatchable to generate power during high demand periods.

Less common types of hydro schemes use water's kinetic energy or undammed sources such as undershot waterwheels.

A simple formula for approximating electric power production at a hydroelectric plant is: $P = hrgk$, where P is Power in kilowatts, h is height in meters, r is flow rate in cubic meters per second, g is acceleration due to gravity of 9.8 m/s^2, and k is a coefficient of efficiency ranging from 0 to 1. Efficiency is often higher with larger and more modern turbines.

Annual electric energy production depends on the available water supply. In some installations the water flow rate can vary by a factor of 10:1 over the course of a year.

INDUSTRIAL HYDROELECTRIC PLANTS

While many hydroelectric projects supply public electricity networks, some are created to serve specific industrial enterprises. Dedicated hydroelectric projects are often built to provide the substantial amounts of electricity needed for aluminium electrolytic plants, for example. In the Scottish Highlands there are examples at Kinlochleven and Lochaber, constructed during the early years of the 20th century. The Grand Coulee Dam, long the world's largest, switched to support

Alcoa aluminum in Bellingham, Washington for America's World War II airplanes before it was allowed to provide irrigation and power to citizens (in addition to aluminum power) after the war. In Suriname, the Brokopondo Reservoir was constructed to provide electricity for the Alcoa aluminium industry. New Zealand's Manapouri Power Station was constructed to supply electricity to the aluminium smelter at Tiwai Point. As of 2007 the Kárahnjúkar Hydropower Project in Iceland remains controversial.

SMALL-SCALE HYDRO-ELECTRIC PLANTS

Although large hydroelectric installations generate most of the world's hydroelectricity, some situations require small hydro plants. These are defined as plants producing up to 10 megawatts, or projects up to 30 megawatts in North America. A small hydro plant may be connected to a distribution grid or may provide power only to an isolated community or a single home.

Small hydro projects generally do not require the protracted economic, engineering and environmental studies associated with large projects, and often can be completed much more quickly. A small hydro development may be installed along with a project for flood control, irrigation or other purposes, providing extra revenue for project costs. In areas that formerly used waterwheels for milling and other purposes, often the site can be redeveloped for electric power production, possibly eliminating the new environmental impact of any demolition operation. Small hydro can be further divided into mini-hydro, units around 1 MW in size, and micro hydro with units as large as 100 kW down to a couple of kW rating.

Small hydro schemes are particularly popular in China, which has over 50% of world small hydro capacity.

Small hydro units in the range 1 MW to about 30 MW are often available from multiple manufacturers using standardized "water to wire" packages; a single contractor can provide all the major mechanical and electrical equipment (turbine, generator, controls, switchgear), selecting from several standard designs to fit the site conditions. Micro hydro projects use a diverse range of equipment; in the smaller sizes industrial centrifugal pumps can be used as turbines, with comparatively low purchase cost compared to purpose-built turbines.

ADVANTAGES

Economics

The major advantage of hydroelectricity is elimination of the cost of fuel. The cost of operating a hydroelectric plant is nearly immune to increases in the cost of fossil fuels such as oil, natural gas or coal, and no imports are needed.

Hydroelectric plants also tend to have longer economic lives than fuel-fired generation, with some plants now in service which were built 50 to 100 years ago. Operating labor cost is also usually low, as plants are automated and have few personnel on site during normal operation.

Where a dam serves multiple purposes, a hydroelectric plant may be added with relatively low construction cost, providing a useful revenue stream to offset the costs of dam operation. It has been calculated that the sale of electricity from the Three Gorges Dam will cover the construction costs after 5 to 8 years of full generation.

Greenhouse Gas Emissions

Since hydroelectric dams do not burn fossil fuels, they do not directly produce carbon dioxide (a greenhouse gas). While some carbon dioxide is produced during manufacture and construction of the project, this is a tiny fraction of the operating emissions of equivalent fossil-fuel electricity generation.

Related Activities

Reservoirs created by hydroelectric schemes often provide facilities for water sports, and become tourist attractions in themselves. In some countries, aquaculture in reservoirs is common. Multi-use dams installed for irrigation support agriculture with a relatively constant water supply. Large hydro dams can control floods, which would otherwise affect people living downstream of the project.

DISADVANTAGES

Environmental Damage

Hydroelectric projects can be disruptive to surrounding aquatic ecosystems both upstream and downstream of the plant site. For instance, studies have shown that dams along the Atlantic and Pacific coasts of North America have reduced salmon populations by preventing access to spawning grounds upstream, even though most dams in salmon habitat have fish ladders installed.

Salmon spawn are also harmed on their migration to sea when they must pass through turbines. This has led to some areas transporting smolt downstream by barge during parts of the year. In some cases dams have been demolished (for example the Marmot Dam demolished in 2007) because of impact on fish. Turbine and power-plant designs that are easier on aquatic life are an active area of research. Mitigation measures such as fish ladders may be required at new projects or as a condition of re-licensing of existing projects.

Generation of hydroelectric power changes the downstream river environment. Water exiting a turbine usually contains very little suspended sediment, which can lead to scouring of river beds and loss of riverbanks. Since turbine gates are often opened intermittently, rapid or even daily fluctuations in river flow are observed.

For example, in the Grand Canyon, the daily cyclic flow variation caused by Glen Canyon Dam was found to be contributing to erosion of sand bars. Dissolved oxygen content of the water may change from pre-construction conditions. Depending on the location, water exiting from turbines is typically much warmer than the pre-dam water, which can change aquatic faunal populations, including endangered species, and prevent natural freezing processes from occurring.

Some hydroelectric projects also use canals to divert a river at a shallower gradient to increase the head of the scheme. In some cases, the entire river may be diverted leaving a dry riverbed. Examples include the Tekapo and Pukaki Rivers.

Greenhouse Gas Emissions

The reservoirs of power plants in tropical regions may produce substantial amounts of methane and carbon dioxide. This is due to plant material in flooded areas decaying in an anaerobic environment, and forming methane, a very potent greenhouse gas. According to the World Commission on Dams report, where the reservoir is large compared to the generating capacity (less than 100 watts per square metre of surface area) and no clearing of the forests in the area was undertaken prior to impoundment of the reservoir, greenhouse gas emissions from the reservoir may be higher than those of a conventional oil-fired thermal generation plant. Although these emissions represent carbon already in the biosphere, not fossil deposits that had been sequestered from the carbon cycle, there is a greater amount of methane due to anaerobic decay, causing greater damage than would otherwise have occurred had the forest decayed naturally. In boreal reservoirs of Canada and Northern Europe, however, greenhouse gas emissions are typically only 2% to 8% of any kind of conventional fossil-fuel thermal generation. A new class of underwater logging operation that targets drowned forests can mitigate the effect of forest decay. Discussions to exclude hydropower facilities from obtaining carbon credits under the Clean Development Mechanism are starting to take place, most recently at the UN Climate Change Conference 2007 in Bali, Indonesia.

Population Relocation

Another disadvantage of hydroelectric dams is the need to relocate the people living where the reservoirs are planned. In February 2008, it was estimated that 40-80 million people worldwide had been physically displaced as a direct result of dam construction. In many cases, no amount of compensation can replace ancestral and cultural attachments to places that have spiritual value to the displaced population. Additionally, historically and culturally important sites can be flooded and lost. Such problems have arisen at the Three Gorges Dam project in China, the Clyde Dam in New Zealand and the Ilýsu Dam in Southeastern Turkey.

Dam Failures

Failures of large dams, while rare, are potentially serious—the Banqiao Dam failure in Southern China resulted in the deaths of 171,000 people and left millions homeless. Dams may be subject to enemy bombardment during wartime, sabotage and terrorism. Smaller dams and micro hydro facilities are less vulnerable to these threats. The creation of a dam in a geologically inappropriate location may cause disasters like the one of the Vajont Dam in Italy, where almost 2000 people died, in 1963.

Affected by Flow Shortage

Changes in the amount of river flow will correlate with the amount of energy produced by a dam. Because of global warming, the volume of glaciers has decreased, such as the North Cascades glaciers, which have lost a third of their volume since 1950, resulting in stream flows that have decreased by as much as 34%. The result of diminished river flow can be power shortages in areas that depend heavily on hydroelectric power.

COMPARISON WITH OTHER METHODS OF POWER GENERATION

Hydroelectricity eliminates the flue gas emissions from fossil fuel combustion, including pollutants such as sulfur dioxide, nitric oxide, carbon monoxide, dust, and mercury in the coal. Hydroelectricity also avoids the hazards of coal mining and the indirect health effects of coal emissions. Compared to nuclear power, hydroelectricity generates no nuclear waste, has none of the dangers associated with uranium mining, nor nuclear leaks. Unlike uranium, hydroelectricity is also a renewable energy source.

Compared to wind farms, hydroelectricity power plants have a more predictable load factor. If the project has a storage reservoir, it can be dispatched to generate power when needed. Hydroelectric plants can be easily regulated to follow variations in power demand.

Unlike fossil-fueled combustion turbines, construction of a hydroelectric plant requires a long lead-time for site studies, hydrological studies, and environmental impact assessment. Hydrological data up to 50 years or more is usually required to determine the best sites and operating regimes for a large hydroelectric plant. Unlike plants operated by fuel, such as fossil or nuclear energy, the number of sites that can be economically developed for hydroelectric production is limited; in many areas the most cost effective sites have already been exploited. New hydro sites tend to be far from population centers and require extensive transmission lines. Hydroelectric generation depends on rainfall in the watershed, and may be significantly reduced in years of low rainfall or snowmelt. Long-term energy yield may be affected by climate change. Utilities that primarily use hydroelectric power may spend additional capital to build extra capacity to ensure sufficient power is available in low water years.

In parts of Canada (the provinces of British Columbia, Manitoba, Ontario, Quebec, Newfoundland and Labrador) hydroelectricity is used so extensively that the word "hydro" is often used to refer to any electricity delivered by a power utility. The government-run power utilities in these provinces are called BC Hydro, Manitoba Hydro, Hydro One (formerly "Ontario Hydro"), Hydro-Québec and Newfoundland and Labrador Hydro respectively. Hydro-Québec is the world's largest hydroelectric generating company, with a total installed capacity (2007) of 35,647 MW, including 33,305 MW of hydroelectric generation.

COUNTRIES WITH THE MOST HYDRO-ELECTRIC CAPACITY

The ranking of hydro-electric capacity is either by actual annual energy production or by installed capacity power rating. A hydro-electric plant rarely operates at its full power rating over a full year; the ratio between annual average power and installed capacity rating is the capacity factor. The installed capacity is the sum of all generator nameplate power ratings. Sources came from *BP Statistical Review - Full Report 2007* List of the largest hydoelectric power stations. Norway produces 98-99% of its electricity from hydroelectric.

Brazil, Canada, Norway and Venezuela are the only countries in the world where the majority of their internal electric energy production is from hydroelectric power.

Country	Annual Hydroelectric Energy Production(TWh)	Installed Capacity (GW)	Capacity Factor	Percent of all electricity
People's Republic of China(2008)	563.3	171.52	0.37	17.18
Brazil	371.5	69.080	0.56	85.56
Canada	368.2	88.974	0.59	61.12
USA	250.8	79.511	0.42	5.74
Russia	179.0	45.000	0.42	17.64
Norway	135.3	27.528	0.49	98.25
India	122.4	33.600	0.43	15.80
Venezuela	83.9	-	-	67.17
Japan	83.6	27.229	0.37	7.21
Sweden	66.2	16.209	0.46	44.34
Paraguay(2006)	64.0	-	-	
France	63.6	25.335	0.25	11.23

OLD HYDRO-ELECTRIC POWER STATIONS

Northern Hemisphere

- Appleton, Wisconsin, USA completed 1882, A waterwheel on the Fox river supplied the first commercial hydroelectric power for lighting to two paper mills and a house, two years after Thomas Edison demonstrated incandescent lighting to the public. Within a matter of weeks of this installation, a power plant was also put into commercial service at Minneapolis.
- Niagara Falls, New York. For many years the largest hydroelectric power station in the world. Operation began locally in 1895 and power was transmitted to Buffalo, New York, in 1896.
- Decew Falls 1, St. Catharines, Ontario, Canada completed 25 August 1898. Owned by Ontario Power Generation. Four units are still operational. Recognized as an IEEE Milestone in Electrical Engineering & Computing by the IEEE Executive Committee in 2002.

- Claverack Creek, in Stottville, New York, believed to be the oldest hydro power site in the United States. The turbine, a Morgan Smith, was constructed in 1869 and installed 2 years later. It is one of the earliest water wheel installations in the United States to generate electricity. It is owned today by Edison Hydro.
- The oldest continuously-operated commercial hydroelectric plant in the United States is built on the Hudson River at Mechanicville, New York. The seven 750 kW units at this station initially supplied power at a frequency of 38 Hz, but later were increased in speed to 40 Hz. It went into commercial service July 22,1898. It is now being restored to its original condition and remains in commercial operation.
- The oldest continuously-operated hydroelectric generator in Canada is located in St. Stephen, New Brunswick, Canada. Part of the construction of the Milltown Cotton Mill, this rope-driven generator originally powered the electric lights for the mill when it opened in 1882, and in 1888 started providing power to homes in the town. NB Power now owns and operates this as part of the Milltown Dam hydroelectric station.

Southern Hemisphere

- A small hydroelectric station, generating 650 kW, opened at Waratah, Tasmania, in 1885
- Duck Reach, Launceston, Tasmania. Completed 1895. The first publicly owned hydro-electric plant in the Southern Hemisphere. Supplied power to the city of Launceston for street lighting.
- Chivilingo was the first hydroelectric plant in Chile and the second in South America. With first power produced in 1897, it has two Pelton wheel turbines each turning a 215 kW generator. It was installed to provide power to mines and the city of Lota, Chile.
- The Snowy Mountains Scheme has turbines all along the tunnel so it, in some perspectives it is also another hydroelectric station. However it only operates during

peak hours of the day and mostly during the evening and early night. It supplies electricity to all over the state of NSW.

MAJOR SCHEMES UNDER CONSTRUCTION

Cost

United States

In the United States, a study is required before constructing a hydroelectric project. In 2008, a study could cost up to $50,000 for a 100 feet (30 m) run of a stream. Both federal and state licenses were required. A license typically cost between $150,000 and $1 million. A project earns money from the sale of energy, the sale of capacity, and the sale of renewable energy credits.

Solar Energy

Solar energy is the radiant light and heat from the Sun that has been harnessed by humans since ancient times using a range of ever-evolving technologies. Solar radiation along with secondary solar resources such as wind and wave power, hydroelectricity and biomass account for most of the available renewable energy on Earth. Only a minuscule fraction of the available solar energy is used.

Solar power technologies provide electrical generation by means of heat engines or photovoltaics. Once converted its uses are only limited by human ingenuity. A partial list of solar applications includes space heating and cooling through solar architecture, potable water via distillation and disinfection, daylighting, hot water, thermal energy for cooking, and high temperature process heat for industrial purposes.

Solar technologies are broadly characterized as either passive solar or active solar depending on the way they capture, convert and distribute sunlight. Active solar techniques include the use of photovoltaic panels, solar thermal collectors, with electrical or mechanical equipment, to convert sunlight into useful outputs. Passive solar techniques include orienting a building to the Sun, selecting materials with favorable thermal mass or light dispersing properties, and designing spaces that naturally circulate air.

Energy from the Sun

The Earth receives 174 petawatts (PW) of incoming solar radiation (insolation) at the upper atmosphere. Approximately 30% is reflected back to space while the rest is absorbed by clouds, oceans and land masses. The spectrum of solar light at the Earth's surface is mostly spread across the visible and near-infrared ranges with a small part in the near-ultraviolet.

Earth's land surface, oceans and atmosphere absorb solar radiation, and this raises their temperature. Warm air containing evaporated water from the oceans rises, causing atmospheric circulation or convection. When the air reaches a high altitude, where the temperature is low, water vapor condenses into clouds, which rain onto the Earth's surface, completing the water cycle.

The latent heat of water condensation amplifies convection, producing atmospheric phenomena such as wind, cyclones and anti-cyclones. Sunlight absorbed by the oceans and land masses keeps the surface at an average temperature of 14 °C. By photosynthesis green plants convert solar energy into chemical energy, which produces food, wood and the biomass from which fossil fuels are derived.

Yearly Solar fluxes & Human Energy Consumption

Solar	3,850,000 EJ
Wind	2,250 EJ
Biomass	3,000 EJ
Primary energy use (2005)	487 EJ
Electricity (2005)	56.7 EJ

The total solar energy absorbed by Earth's atmosphere, oceans and land masses is approximately 3,850,000 exajoules (EJ) per year. In 2002, this was more energy in one hour than the world used in one year. Photosynthesis captures approximately 3,000 EJ per year in biomass. The amount of solar energy reaching the surface of the planet is so vast that in one year it is about twice as much as will ever be obtained from all of the Earth's non-renewable resources of coal, oil, natural gas, and mined uranium combined.

From the table of resources it would appear that solar, wind or biomass would be sufficient to supply all of our energy needs, however, the increased use of biomass has had a negative effect on global warming and dramatically increased food prices by diverting forests and crops into biofuel production. As intermittent resources, solar and wind raise other issues.

APPLICATIONS OF SOLAR TECHNOLOGY

Solar energy refers primarily to the use of solar radiation for practical ends. All other renewable energies other than geothermal derive their energy from the sun.

Solar technologies are broadly characterized as either passive or active depending on the way they capture, convert and distribute sunlight. Active solar techniques use photovoltaic panels, pumps, and fans to convert sunlight into useful outputs. Passive solar techniques include selecting materials with favorable thermal properties, designing spaces that naturally circulate air, and referencing the position of a building to the Sun. Active solar technologies increase the supply of energy and are considered supply side technologies, while passive solar technologies reduce the need for alternate resources and are generally considered demand side technologies.

ARCHITECTURE AND URBAN PLANNING

Sunlight has influenced building design since the beginning of architectural history. Advanced solar architecture and urban planning methods were first employed by the Greeks and Chinese, who oriented their buildings toward the south to provide light and warmth.

The common features of passive solar architecture are orientation relative to the Sun, compact proportion (a low surface area to volume ratio), selective shading (overhangs) and thermal mass. When these features are tailored to the local climate and environment they can produce well-lit spaces that stay in a comfortable temperature range. Socrates' Megaron House is a classic example of passive solar design. The most recent approaches to solar design use computer modeling tying together solar lighting, heating and ventilation systems in an integrated solar design package. Active solar equipment such as pumps, fans and switchable windows can complement passive design and improve system performance.

Urban heat islands (UHI) are metropolitan areas with higher temperatures than that of the surrounding environment. The higher temperatures are a result of increased absorption of the Solar light by urban materials such as asphalt and concrete, which have lower albedos and higher heat capacities than those in the natural environment. A straightforward method of counteracting the UHI effect is to paint buildings and roads white and plant trees. Using these methods, a hypothetical "cool communities" program in Los Angeles has projected that urban temperatures could be reduced by approximately 3 °C at an estimated cost of US$1 billion, giving estimated total annual benefits of US$530 million from reduced air-conditioning costs and healthcare savings.

AGRICULTURE AND HORTICULTURE

Agriculture seeks to optimize the capture of solar energy in order to optimize the productivity of plants. Techniques such as timed planting cycles, tailored row orientation, staggered heights between rows and the mixing of plant varieties can improve crop yields. While sunlight is generally considered a plentiful resource, the exceptions highlight the importance of solar energy to agriculture. During the short growing seasons of the Little Ice Age, French and English farmers employed fruit walls to maximize the collection of solar energy.

These walls acted as thermal masses and accelerated ripening by keeping plants warm. Early fruit walls were built perpendicular to the ground and facing south, but over time, sloping walls were developed to make better use of sunlight. In 1699, Nicolas Fatio de Duillier even suggested using a tracking mechanism which could pivot to follow the Sun. Applications of solar energy in agriculture aside from growing crops include pumping water, drying crops, brooding chicks and drying chicken manure. More recently the technology has been embraced by vinters, who use the energy generated by solar panels to power grape presses.

Greenhouses convert solar light to heat, enabling year-round production and the growth (in enclosed environments) of specialty crops and other plants not naturally suited to the local climate. Primitive greenhouses were first used during Roman times to produce cucumbers year-round for the Roman emperor Tiberius. The first modern greenhouses were built in Europe in the 16th century to keep exotic plants brought back from explorations abroad. Greenhouses remain an important part of horticulture today, and plastic transparent materials have also been used to similar effect in polytunnels and row covers.

Solar Lighting

The history of lighting is dominated by the use of natural light. The Romans recognized a right to light as early as the 6th century and English law echoed these judgments with the Prescription Act of 1832. In the 20th century artificial lighting became the main source of interior illumination but daylighting techniques and hybrid solar lighting solutions are ways to reduce energy consumption.

Daylighting systems collect and distribute sunlight to provide interior illumination. This passive technology directly offsets energy use by replacing artificial lighting, and indirectly offsets non-solar energy use by reducing the need for air-conditioning. Although difficult to quantify, the use of natural lighting also offers physiological and psychological benefits compared to artificial lighting. Daylighting design implies careful selection of window types, sizes and orientation; exterior shading devices may be considered as well. Individual features include sawtooth roofs, clerestory windows, light shelves, skylights and light tubes. They may be incorporated into existing structures, but are most effective when integrated into a solar design package that accounts for factors such as glare, heat flux and time-of-use. When daylighting features are properly implemented they can reduce lighting-related energy requirements by 25%.

Hybrid solar lighting is an active solar method of providing interior illumination. HSL systems collect sunlight using focusing mirrors that track the Sun and use optical fibers to transmit it inside the building to supplement conventional lighting. In single-story applications these systems are able to transmit 50% of the direct sunlight received. Solar lights that charge during the day and light up at dusk are a common sight along walkways. Although daylight saving time is promoted as a way to use sunlight to save energy, recent research has been limited and reports contradictory results: several studies report savings, but just as many suggest no effect or even a net loss, particularly when gasoline consumption is taken into account. Electricity use is greatly affected by geography, climate and economics, making it hard to generalize from single studies.

Solar Thermal

Solar thermal technologies can be used for water heating, space heating, space cooling and process heat generation.

Water Heating

Solar hot water systems use sunlight to heat water. In low geographical latitudes (below 40 degrees) from 60 to 70% of the domestic hot water use with temperatures up to 60 °C can be provided by solar heating systems. The most common types of solar water heaters are evacuated tube collectors (44%) and glazed flat plate collectors (34%) generally used for domestic hot water; and unglazed plastic collectors (21%) used mainly to heat swimming pools.

As of 2007, the total installed capacity of solar hot water systems is approximately 154 GW. China is the world leader in their deployment with 70 GW installed as of 2006 and a long term goal of 210 GW by 2020. Israel and Cyprus are the per capita leaders in the use of solar hot water systems with over 90% of homes using them. In the United States, Canada and Australia heating swimming pools is the dominant application of solar hot water with an installed capacity of 18 GW as of 2005.

HEATING, COOLING AND VENTILATION

In the United States, heating, ventilation and air conditioning (HVAC) systems account for 30% (4.65 EJ) of the energy used in commercial buildings and nearly 50% (10.1 EJ) of the energy used in residential buildings. Solar heating, cooling and ventilation technologies can be used to offset a portion of this energy.

Thermal mass is any material that can be used to store heat—heat from the Sun in the case of solar energy. Common thermal mass materials include stone, cement and water. Historically they have been used in arid climates or warm temperate regions to keep buildings cool by absorbing solar energy during the day and radiating stored heat to the cooler atmosphere at night. However they can be used in cold temperate areas to maintain warmth as well.

The size and placement of thermal mass depend on several factors such as climate, daylighting and shading conditions. When properly incorporated, thermal mass maintains space temperatures in a comfortable range and reduces the need for auxiliary heating and cooling equipment.

A solar chimney (or thermal chimney, in this context) is a passive solar ventilation system composed of a vertical shaft connecting the interior and exterior of a building. As the chimney warms, the air inside is heated causing an updraft that pulls air through the building. Performance can be improved by using glazing and thermal mass materials in a way that mimics greenhouses.

Deciduous trees and plants have been promoted as a means of controlling solar heating and cooling. When planted on the southern side of a building, their leaves provide shade during the summer, while the bare limbs allow light to pass during the winter.

Since bare, leafless trees shade 1/3 to 1/2 of incident solar radiation, there is a balance between the benefits of summer shading and the corresponding loss of winter heating. In climates with significant heating loads, deciduous trees should not be planted on the southern side of a building because they will interfere with winter solar availability. They can, however, be used on the east and west sides to provide a degree of summer shading without appreciably affecting winter solar gain.

Water Treatment

Solar distillation can be used to make saline or brackish water potable. The first recorded instance of this was by 16th century Arab alchemists. A large-scale solar distillation project was first constructed in 1872 in the Chilean mining town of Las Salinas. The plant, which had solar collection area of 4,700 m^2, could produce up to 22,700 L per day and operated for 40 years. Individual still designs include single-slope, double-slope (or greenhouse type), vertical, conical, inverted absorber, multi-wick, and multiple effect. These stills can operate in passive, active, or hybrid modes. Double-slope stills are the most economical for decentralized domestic purposes, while active multiple effect units are more suitable for large-scale applications.

Solar water disinfection (SODIS) involves exposing water-filled plastic polyethylene terephthalate (PET) bottles to sunlight for several hours. Exposure times vary depending on weather and climate from a minimum of six hours to two days during fully overcast conditions. SODIS is recommended by the World Health Organization as a viable method for household water treatment and safe storage. Over two million people in developing countries use SODIS for their daily drinking water.

Solar energy may be used in a water stabilisation pond to treat waste water without chemicals or electricity. A further environmental advantage is that algae grow in such ponds and consume carbon dioxide in photosynthesis.

Solar energy has also be used for detoxification of contaminated water via photolysis. The economics of this process is though doubtful.

Cooking

Solar cookers use sunlight for cooking, drying and pasteurization. They can be grouped into three broad categories: box cookers, panel cookers and reflector cookers. The simplest solar cooker—the box cooker first built by Horace de Saussure in 1767. A basic box cooker consists of an insulated container with a transparent lid. It can be used effectively with partially overcast skies and will typically reach temperatures of 90–150 °C. Panel cookers use a reflective panel to direct sunlight onto an insulated container and reach temperatures comparable to box cookers. Reflector cookers use various concentrating geometries (dish, trough, Fresnel mirrors) to focus light on a cooking container. These cookers reach temperatures of 315 °C and above but require direct light to function properly and must be repositioned to track the Sun.

The solar bowl is a concentrating technology employed by the Solar Kitchen in Auroville, India, where a stationary spherical reflector focuses light along a line perpendicular to the sphere's interior surface, and a computer control system moves the receiver to intersect this line. Steam is produced in the receiver at temperatures reaching 150 °C and then used for process heat in the kitchen.

A reflector developed by Wolfgang Scheffler in 1986 is used in many solar kitchens. Scheffler reflectors are flexible parabolic dishes that combine aspects of trough and power tower concentrators. Polar tracking is used to follow the Sun's daily course and the curvature of the reflector is adjusted for seasonal variations in the incident angle of sunlight. These reflectors can reach temperatures of 450–650 °C and have a fixed focal point, which simplifies cooking. The world's largest Scheffler reflector system in Abu Road, Rajasthan, India is capable of cooking up to 35,000 meals a day. As of 2008, over 2,000 large Scheffler cookers had been built worldwide.

Process Heat

Solar concentrating technologies such as parabolic dish, trough and Scheffler reflectors can provide process heat for commercial and industrial applications. The first commercial system was the Solar Total Energy Project (STEP) in Shenandoah, Georgia, USA where a field of 114 parabolic dishes provided 50% of the process heating, air conditioning and electrical requirements for a clothing factory. This grid-connected cogeneration system provided 400 kW of electricity plus thermal energy in the form of 401 kW steam and 468 kW chilled water, and had a one hour peak load thermal storage. Evaporation ponds are shallow pools that concentrate dissolved solids through evaporation. The use of evaporation ponds to obtain salt from sea water is one of the oldest applications of solar energy. Modern uses include concentrating brine solutions used in leach mining and removing dissolved solids from waste streams.

Clothes lines, clotheshorses, and clothes racks dry clothes through evaporation by wind and sunlight without consuming electricity or gas. In some states of the United States legislation protects the "right to dry" clothes.

Unglazed transpired collectors (UTC) are perforated sun-facing walls used for preheating ventilation air. UTCs can raise the incoming air temperature up to 22 °C and deliver outlet temperatures of 45–60 °C. The short payback period of transpired collectors (3 to 12 years) makes them a more cost-effective alternative than glazed collection systems. As of 2003, over 80 systems with a combined collector area of 35,000 m^2 had been installed worldwide, including an 860 m^2 collector in Costa Rica used for drying coffee beans and a 1,300 m^2 collector in Coimbatore, India used for drying marigolds.

Electrical Generation

Sunlight can be converted into electricity using photovoltaics (PV), concentrating solar power (CSP), and various experimental technologies. PV has mainly been used to power small and medium-sized applications, from the calculator powered by a single solar cell to off-grid homes powered by a photovoltaic array. For large-scale generation, CSP plants like SEGS have been the norm but recently multi-megawatt PV plants are becoming common. Completed in 2007, the 14 MW power station in Clark County, Nevada and the 20 MW site in Beneixama, Spain are characteristic of the trend toward larger photovoltaic power stations in the US and Europe. As an intermittent power source, solar power requires a backup supply, which can partially be complemented with wind power. Local backup usually is done with batteries, while utilities normally use pumped-hydro storage. The Institute for Solar Energy Supply Technology of the University of Kassel pilot-tested a combined power plant linking solar, wind, biogas and hydrostorage to provide load-following power around the clock, entirely from renewable sources.

Photovoltaics

A solar cell, or photovoltaic cell (PV), is a device that converts light into direct current using the photoelectric effect. The first solar cell was constructed by Charles Fritts in the 1880s. Although the prototype selenium cells converted less than 1% of incident light into electricity, both Ernst Werner von Siemens and James Clerk Maxwell recognized the importance of this discovery. Following the work of Russell Ohl in the 1940s, researchers Gerald Pearson, Calvin Fuller and Daryl Chapin created the silicon solar cell in 1954. These early solar cells cost 286 USD/watt and reached efficiencies of 4.5–6%.

The earliest significant application of solar cells was as a back-up power source to the Vanguard I satellite in 1958, which allowed it to continue transmitting for over a year after its chemical battery was exhausted. The successful operation of solar cells on this mission was duplicated in many other Soviet and American satellites, and by the late 1960s, PV had become the established source of power for them. Photovoltaics went on to play an essential part in the success of early commercial satellites such as Telstar, and they remain vital to the telecommunications infrastructure today.

The high cost of solar cells limited terrestrial uses throughout the 1960s. This changed in the early 1970s when prices reached levels that made PV generation competitive in remote areas without grid access. Early terrestrial uses included powering telecommunication stations, off-shore oil rigs, navigational buoys and railroad crossings. These off-grid applications have proven very successful and accounted for over half of worldwide installed capacity until 2004.

The 1973 oil crisis stimulated a rapid rise in the production of PV during the 1970s and early 1980s. Economies of scale which resulted from increasing production along with improvements in system performance brought the price of PV down from 100 USD/watt in 1971 to 7 USD/watt in 1985.

Steadily falling oil prices during the early 1980s led to a reduction in funding for photovoltaic R&D and a discontinuation of the tax credits associated with the Energy Tax Act of 1978. These factors moderated growth to approximately 15% per year from 1984 through 1996.

Since the mid-1990s, leadership in the PV sector has shifted from the US to Japan and Germany. Between 1992 and 1994 Japan increased R&D funding, established net metering guidelines, and introduced a subsidy program to encourage the installation of residential PV systems. As a result, PV installations in the country climbed from 31.2 MW in 1994 to 318 MW in 1999, and worldwide production growth increased to 30% in the late 1990s.

Germany has become the leading PV market worldwide since revising its Feed-in tariff system as part of the Renewable Energy Sources Act. Installed PV capacity has risen from 100 MW in 2000 to approximately 4,150 MW at the end of 2007. Spain has become the third largest PV market after adopting a similar feed-in tariff structure in 2004, while France, Italy, South Korea and the US have seen rapid growth recently due to various incentive programs and local market conditions. The power output of domestic photovoltaic devices is usually described in kilowatt-peak (kWp) units.

CONCENTRATING SOLAR POWER

Concentrated sunlight has been used to perform useful tasks since the time of ancient China. A legend claims that Archimedes used polished shields to concentrate sunlight on the invading Roman fleet and repel them from Syracuse. Auguste Mouchout used a parabolic trough to produce steam for the first solar steam engine in 1866, and subsequent developments led to the use of concentrating solar-powered devices for irrigation, refrigeration and locomotion.

Concentrating Solar Power (CSP) systems use lenses or mirrors and tracking systems to focus a large area of sunlight into a small beam. The concentrated light is then used as a heat source for a conventional power plant. A wide range of concentrating technologies exists; the most developed are the solar trough, parabolic dish and solar power tower. These methods vary in the way they track the Sun and focus light. In all these systems a working fluid is heated by the concentrated sunlight, and is then used for power generation or energy storage.

A solar trough consists of a linear parabolic reflector that concentrates light onto a receiver positioned along the reflector's focal line. The reflector is made to follow the Sun during the daylight hours by tracking along a single axis. Trough systems provide the best land-use factor of any solar technology. The SEGS plants in California and Acciona's Nevada Solar One near Boulder City, Nevada are representatives of this technology.

A parabolic dish system consists of a stand-alone parabolic reflector that concentrates light onto a receiver positioned at the reflector's focal point. The reflector tracks the Sun along two axes. Parabolic dish systems give the highest efficiency among CSP technologies. The 50 kW Big Dish in Canberra, Australia is an example of this technology. The stirling solar dish combines a parabolic concentrating dish with a stirling heat engine which normally drives an electric generator. The advantages of stirling solar over photovoltaic cells are higher efficiency of converting sunlight into electricity and longer lifetime. A stirling engine has an approximate mean time before failure (MTBF) of 25 years.

A solar power tower uses an array of tracking reflectors (heliostats) to concentrate light on a central receiver atop a tower. Power towers are less advanced than trough systems but offer higher efficiency and better energy storage capability. The Solar Two in Barstow, California and the Planta Solar 10 in Sanlucar la Mayor, Spain are representatives of this technology.

EXPERIMENTAL SOLAR POWER

A solar updraft tower (also known as a solar chimney or solar tower) consists of a large greenhouse that funnels into a central tower. As sunlight shines on the greenhouse, the air inside is heated, and expands. The expanding air flows toward the central tower, where a turbine converts the air flow into electricity. A 50 kW prototype was constructed in Ciudad Real, Spain and operated for eight years before decommissioning in 1989.

A solar pond is a pool of salt water (usually 1–2 m deep) that collects and stores solar energy. Solar ponds were first proposed by Dr. Rudolph Bloch in 1948 after he came across reports of a lake in Hungary in which the temperature increased with depth. This effect was due to salts in the lake's water, which created a "density gradient" that prevented convection currents. A prototype was constructed in 1958 on the shores of the Dead Sea near Jerusalem. The pond consisted of layers of water that successively increased from a weak salt solution at the top to a high salt solution at the bottom. This solar pond was capable of producing temperatures of 90 °C in its bottom layer and had an estimated solar-to-electric efficiency of two percent.

Thermoelectric, or "thermovoltaic" devices convert a temperature difference between dissimilar materials into an electric current. First proposed as a method to store solar energy by solar pioneer Mouchout in the 1800s, thermoelectrics reemerged in the Soviet Union during the 1930s. Under the direction of Soviet scientist Abram Ioffe a concentrating system was used to thermoelectrically generate power for a 1 hp engine. Thermogenerators were later used in the US space program as an energy conversion technology for powering deep space missions such as Cassini, Galileo and Viking. Research in this area is focused on raising the efficiency of these devices from 7–8% to 15–20%.

Space solar power systems would use a large solar array in geosynchronous orbit to collect sunlight and beam this energy in the form of microwave radiation to receivers (rectennas) on Earth for distribution. This concept was first proposed by Dr. Peter Glaser in 1968 and since then a wide variety of systems have been studied with both photovoltaic and concentrating solar thermal technologies being proposed. Although still in the concept stage, these systems offer the possibility of delivering power approximately 96% of the time. In 2008, John C. Mankins, a former NASA scientist, successfully used radio waves to send solar power between two Hawaiian islands in an experiment funded by the Discovery Channel. Mankins claims that this "proves the technology exists to beam solar power from satellites back to Earth."

Solar Chemical

Solar chemical processes use solar energy to drive chemical reactions. These processes offset energy that would otherwise come from an alternate source and can convert solar energy into storable and transportable fuels. Solar induced chemical reactions can be divided into thermochemical or photochemical.

Hydrogen production technologies been a significant area of solar chemical research since the 1970s. Aside from electrolysis driven by photovoltaic or photochemical cells, several thermochemical processes have also been explored. One such route uses concentrators to split water into oxygen and hydrogen at high temperatures (2300-2600 °C). Another approach uses the heat from solar concentrators to drive the steam reformation of natural gas thereby increasing the overall hydrogen yield compared to conventional reforming methods. Thermochemical cycles characterized by the decomposition and regeneration of reactants present another avenue for hydrogen production. The Solzinc process under development at the Weizmann Institute uses a 1 MW solar furnace to decompose zinc oxide (ZnO) at temperatures above 1200 °C. This initial reaction produces pure zinc, which can subsequently be reacted with water to produce hydrogen.

Sandia's Sunshine to Petrol (S2P) technology uses the high temperatures generated by concentrating sunlight along with a zirconia/ferrite catalyst to break down atmospheric carbon dioxide into oxygen and carbon monoxide (CO). The carbon monoxide can then be used to synthesize conventional fuels such as methanol, gasoline and jet fuel.

A photogalvanic device is a type of battery in which the cell solution (or equivalent) forms energy-rich chemical intermediates when illuminated. These energy-rich intermediates can potentially be stored and subsequently reacted at the electrodes to produce an electric potential. The ferric-thionine chemical cell is an example of this technology.

Photoelectrochemical cells or PECs consist of a semiconductor, typically titanium dioxide or related titanates, immersed in an electrolyte. When the semiconductor is illuminated an electrical potential develops. There are two types of photoelectrochemical cells: photoelectric cells that convert light into electricity and photochemical cells that use light to drive chemical reactions such as electrolysis.

Solar Vehicles

Development of a solar powered car has been an engineering goal since the 1980s. The World Solar Challenge is a biannual solar-powered car race, where teams from universities and enterprises compete over 3,021 kilometres (1,877 mi) across central Australia from Darwin to Adelaide. In 1987, when it was founded, the winner's average speed was 67 kilometres per hour (42 mph) and by 2007 the winner's average speed had improved to 90.87 kilometres per hour (56.46 mph). The North American Solar Challenge and the planned South African Solar Challenge are comparable competitions that reflect an international interest in the engineering and development of solar powered vehicles.

Some vehicles use solar panels for auxiliary power, such as for air conditioning, to keep the interior cool, thus reducing fuel consumption.

In 1975, the first practical solar boat was constructed in England. By 1995, passenger boats incorporating PV panels began appearing and are now used extensively. In 1996, Kenichi Horie made the first solar powered crossing of the Pacific Ocean, and the *sun21* catamaran made the first solar powered crossing of the Atlantic Ocean in the winter of 2006–2007. There are plans to circumnavigate the globe in 2010.

In 1974, the unmanned *Sunrise II* plane made the first solar flight. On 29 April 1979, the *Solar Riser* made the first flight in a solar powered, fully controlled, man carrying flying machine, reaching an altitude of 40 feet (12 m). In 1980, the *Gossamer Penguin* made the first piloted flights powered solely by photovoltaics. This was quickly followed by the *Solar Challenger* which crossed the English Channel in July 1981. In 1990 Eric Raymond in 21 hops flew from California to North Carolina using solar power. Developments then turned back to unmanned aerial vehicles (UAV) with the *Pathfinder* (1997) and subsequent designs, culminating in the *Helios* which set the altitude record for a non-rocket-propelled aircraft at 29,524 metres (96,860 ft) in 2001. The *Zephyr*, developed by BAE Systems, is the latest in a line of record-breaking solar aircraft, making a 54-hour flight in 2007, and month-long flights are envisioned by 2010.

A solar balloon is a black balloon that is filled with ordinary air. As sunlight shines on the balloon, the air inside is heated and expands causing an upward buoyancy force, much like an artificially heated hot air balloon. Some solar balloons are large enough for human flight, but usage is generally limited to the toy market as the surface-area to payload-weight ratio is relatively high.

Solar sails are a proposed form of spacecraft propulsion using large membrane mirrors to exploit radiation pressure from the Sun. Unlike rockets, solar sails require no fuel. Although the thrust is small compared to rockets, it continues as long as the Sun shines onto the deployed sail and in the vacuum of space significant speeds can eventually be achieved.

The High-altitude airship (HAA) is an unmanned, long-duration, lighter-than-air vehicle using helium gas for lift, and thin-film solar cells for power. The United States Department of Defense Missile Defense Agency has contracted Lockheed Martin to construct it to enhance the Ballistic Missile Defense System (BMDS). Airships have some advantages for solar-powered flight: they do not require power to remain aloft, and an airship's envelope presents a large area to the Sun.

ENERGY STORAGE METHODS

Solar energy is not available at night, and energy storage is an important issue because modern energy systems usually assume continuous availability of energy. Thermal mass systems can store solar energy in the form of heat at domestically useful temperatures for daily or seasonal durations. Thermal storage systems generally use readily available materials with high specific heat capacities such as water, earth and stone. Well-designed systems can lower peak demand, shift time-of-use to off-peak hours and reduce overall heating and cooling requirements.

Phase change materials such as paraffin wax and Glauber's salt are another thermal storage media. These materials are inexpensive, readily available, and can deliver domestically useful temperatures (approximately 64 °C). The "Dover House" (in Dover, Massachusetts) was the first to use a Glauber's salt heating system, in 1948.

Solar energy can be stored at high temperatures using molten salts. Salts are an effective storage medium because they are low-cost, have a high specific heat capacity and can deliver heat at temperatures compatible with conventional power systems. The Solar Two used this method of energy storage, allowing it to store 1.44 TJ in its 68 m^3 storage tank with an annual storage efficiency of about 99%.

Off-grid PV systems have traditionally used rechargeable batteries to store excess electricity. With grid-tied systems, excess electricity can be sent to the transmission grid. Net metering programs give these systems a credit for the electricity they deliver to the grid. This credit offsets electricity provided from the grid when the system cannot meet demand, effectively using the grid as a storage mechanism.

Pumped-storage hydroelectricity stores energy in the form of water pumped when energy is available from a lower elevation reservoir to a higher elevation one. The energy is recovered when demand is high by releasing the water to run through a hydroelectric power generator.

DEVELOPMENT, DEPLOYMENT AND ECONOMICS

Beginning with the surge in coal use which accompanied the Industrial Revolution, energy consumption has steadily transitioned from wood and biomass to fossil fuels. The early development of solar technologies starting in the 1860s was driven by an expectation that coal would soon become scarce.

However development of solar technologies stagnated in the early 20th century in the face of the increasing availability, economy, and utility of coal and petroleum.

The 1973 oil embargo and 1979 energy crisis caused a reorganization of energy policies around the world and brought renewed attention to developing solar technologies. Deployment strategies focused on incentive programs such as the Federal Photovoltaic Utilization Program in the US and the Sunshine Program in Japan. Other efforts included the formation of research facilities in the US (SERI, now NREL), Japan (NEDO), and Germany (Fraunhofer Institute for Solar Energy Systems ISE).

Between 1970 and 1983 photovoltaic installations grew rapidly, but falling oil prices in the early 1980s moderated the growth of PV from 1984 to 1996. Since 1997, PV development has accelerated due to supply issues with oil and natural gas, global warming concerns, and the improving economic position of PV relative to other energy technologies. Photovoltaic production growth has averaged 40% per year since 2000 and installed capacity reached 10.6 GW at the end of 2007. Since 2006 it has been economical for investors to install photovoltaics for free in return for a long term power purchase agreement. 50% of commercial systems were installed in this manner in 2007 and it is expected that 90% will by 2009. Nellis Air Force Base is receiving photoelectric power for about 2.2 ¢/kWh and grid power for 9 ¢/kWh.

Commercial solar water heaters began appearing in the United States in the 1890s. These systems saw increasing use until the 1920s but were gradually replaced by cheaper and more reliable heating fuels. As with photovoltaics, solar water heating attracted renewed attention as a result of the oil crises in the 1970s but interest subsided in the 1980s due to falling petroleum prices. Development in the solar water heating sector progressed steadily throughout the 1990s and growth rates have averaged 20% per year since 1999. Although generally underestimated, solar water heating is by far the most widely deployed solar technology with an estimated capacity of 154 GW as of 2007.

Commercial concentrating solar power (CSP) plants were first developed in the 1980s. CSP plants such as SEGS project in the United States have a LEC of 12–14 ¢/kWh. The 11 MW PS10 power tower in Spain, completed in late 2005, is Europe's first commercial CSP system, and a total capacity of 300 MW is expected to be installed in the same area by 2013.

Solar installations in recent years have also largely begun to expand into residential areas, with governments offering incentive programs to make "green" energy a more economically viable option. In Canada the government offers the RESOP (Renewable Energy Standard Offer Program). The program allows residential homeowners with solar panel installations to sell the energy they produce back to the grid (i.e., the government) at 41¢/kWh, while drawing power from the grid at an average rate of 20¢/kWh. The program is designed to help promote the government's green agenda and lower the strain often placed on the energy grid at peak hours. With the incentives offered by the program the average payback period for a residential solar installation (sized between 1.3 kW and 5 kW) is estimated at 18 to 23 years, considering such cost factors as parts, installation and maintenance, as well as the average energy production of a system on an annual basis.

Daniel Lincot, the chairman of the 2008 European Photovoltaic Solar Energy Conference and the research director of the Paris-based Photovoltaic Energy Development and Research Institute, said that photovoltaics can cover all the world energy demand. Photovoltaics are 85 times as efficient as growing corn for ethanol. On a 300 feet (91 m) by 300 feet (1 hectare) plot of land enough ethanol can be produced to drive a car 30,000 miles (48,000 km) per year or 2,500,000 miles (4,020,000 km) by covering the same land with photo cells. The deserts of the South Western United States could produce sufficient electricity to fulfill all of the electrical needs of the United States, and could use electrolysis to produce Hydrogen from water to power aircraft.

3

Tidal Power Energy Resources

Tidal power, sometimes called tidal energy, is a form of hydropower that converts the energy of tides into electricity or other useful forms of power. Although not yet widely used, tidal power has potential for future electricity generation. Tides are more predictable than wind energy and solar power. Historically, tide mills have been used, both in Europe and on the Atlantic coast of the USA. The earliest occurrences date from the Middle Ages, or even from Roman times.

GENERATION OF TIDAL ENERGY

Tidal power is the only form of energy which derives directly from the relative motions of the Earth–Moon system, and to a lesser extent from the Earth–Sun system. The tidal forces produced by the Moon and Sun, in combination with Earth's rotation, are responsible for the generation of the tides. Other sources of energy originate directly or indirectly from the Sun, including fossil fuels, conventional hydroelectric, wind, biofuels, wave power and solar. Nuclear is derived using radioactive material from the Earth, geothermal power uses the heat of magma below the Earth's crust, which comes from radioactive decay.

Tidal energy is generated by the relative motion of the Earth, Sun and the Moon, which interact via gravitational forces. Periodic changes of water levels, and associated tidal currents, are due to the gravitational attraction by the Sun and Moon. The magnitude of the tide at a location is the result of the changing positions of the Moon and Sun relative to the

Earth, the effects of Earth rotation, and the local shape of the sea floor and coastlines.

Because the Earth's tides are caused by the tidal forces due to gravitational interaction with the Moon and Sun, and the Earth's rotation, tidal power is practically inexhaustible and classified as a renewable energy source. A tidal energy generator uses this phenomenon to generate energy. The stronger the tide, either in water level height or tidal current velocities, the greater the potential for tidal energy generation.

Tidal movement causes a continual loss of mechanical energy in the Earth–Moon system due to pumping of water through the natural restrictions around coastlines, and due to viscous dissipation at the seabed and in turbulence. This loss of energy has caused the rotation of the Earth to slow in the 4.5 billion years since formation. During the last 620 million years the period of rotation has increased from 21.9 hours to the 24 hours we see now; in this period the Earth has lost 17% of its rotational energy. While tidal power may take additional energy from the system, increasing the rate of slowdown, the effect would be noticeable over millions of years only, thus being negligible.

CATEGORIES OF TIDAL POWER

Tidal power can be classified into two main types:

- Tidal stream systems make use of the kinetic energy of moving water to power turbines, in a similar way to windmills that use moving air. This method is gaining in popularity because of the lower cost and lower ecological impact compared to barrages.
- Barrages make use of the potential energy in the difference in height (or *head*) between high and low tides. Barrages are essentially dams across the full width of a tidal estuary, and suffer from very high civil infrastructure costs, a worldwide shortage of viable sites, and environmental issues.
- Tidal lagoons, are similar to barrages, but can be constructed as self contained structures, not fully across an estuary, and are claimed to incur much lower cost

and impact overall. Furthermore they can be configured to generate continuously which is not the case with barrages.

Modern advances in turbine technology may eventually see large amounts of power generated from the ocean, especially tidal currents using the tidal stream designs but also from the major thermal current systems such as the Gulf Stream, which is covered by the more general term marine current power. Tidal stream turbines may be arrayed in high-velocity areas where natural tidal current flows are concentrated such as the west and east coasts of Canada, the Strait of Gibraltar, the Bosporus, and numerous sites in south east Asia and Australia. Such flows occur almost anywhere where there are entrances to bays and rivers, or between land masses where water currents are concentrated.

Tidal Stream Generators

A relatively new technology, tidal stream generators draw energy from currents in much the same way as wind turbines. The higher density of water, 832 times the density of air, means that a single generator can provide significant power at low tidal flow velocities (compared with wind speed). Given that power varies with the density of medium and the cube of velocity, it is simple to see that water speeds of nearly one-tenth of the speed of wind provide the same power for the same size of turbine system. However this limits the application in practice to places where the tide moves at speeds of at least 2 knots (1m/s) even close to neap tides.

Since tidal stream generators are an immature technology (no commercial scale production facilities are yet routinely supplying power), no standard technology has yet emerged as the clear winner, but a large variety of designs are being experimented with, some very close to large scale deployment. Several prototypes have shown promise with many companies making bold claims, some of which are yet to be independently verified, but they have not operated commercially for extended periods to establish performances and rates of return on investments.

Engineering Approaches

The European Marine Energy Centre categorises them under four heads although a number of other approaches are also being tried.

Horizontal Axis Turbines

These are close in concept to traditional windmills operating under the sea and have the most prototypes currently operating. These include:

Kvalsund, south of Hammerfest, Norway. Although still a prototype, a turbine with a reported capacity of 300 kW was connected to the grid on 13 November 2003. A 300 kW Periodflow marine current propeller type turbine — Seaflow — was installed by Marine Current Turbines off the coast of Lynmouth, Devon, England, in 2003. The 11m diameter turbine generator was fitted to a steel pile which was driven into the seabed. As a prototype, it was connected to a dump load, not to the grid.

Since April 2007 Verdant Power has been running a prototype project in the East River between Queens and Roosevelt Island in New York City; it was the first major tidal-power project in the United States. The strong currents pose challenges to the design: the blades of the 2006 and 2007 prototypes broke off, and new reinforced turbines were installed in September 2008.

Following the Seaflow trial, a fullsize prototype, called SeaGen, was installed by Marine Current Turbines in Strangford Lough in Northern Ireland in April 2008. The turbine began to generate at full power of just over 1.2 MW in December 2008 and was reported to have fed 150kW into the grid for the first time on 17 July 2008. It is currently the only commercial scale device to have been installed anywhere in the world. OpenHydro, an Irish company exploiting the Open-Centre Turbine developed in the U.S., has a prototype being tested at the European Marine Energy Centre (EMEC), in Orkney, Scotland.

Vertical Axis Turbines

The Gorlov turbine is a variant of the Darrieus design featuring a helical design which is being commercially piloted on a large scale in S. Korea. Neptune Renewable Energy has developed Proteus which uses a barrage of vertical axis crossflow turbines for use mainly in estuaries.

Oscillating Devices

These don't use rotary devices at all but rather aerofoil sections which are pushed sideways by the flow.

Oscillating stream power extraction was proven with the omni or bi-directional Wing'd Pump windmill

During 2003 a 150kW oscillating hydroplane device, the Stingray, was tested off the Scottish coast.

Venturi Effect

This uses a shroud to increase the flow rate through the turbine. These can be mounted horizontally or vertically.

The Australian company Tidal Energy Pty Ltd undertook successful commercial trials of highly efficient shrouded tidal turbines on the Gold Coast, Queensland in 2002. Tidal Energy has commenced a rollout of their shrouded turbine for a remote Australian community in northern Australia where there are some of the fastest flows ever recorded (11 m/s, 21 knots) – two small turbines will provide 3.5 MW. Another larger 5 meter diameter turbine, capable of 800 kW in 4 m/s of flow, is planned for deployment as a tidal powered desalination showcase near Brisbane Australia in October 2008. Another device, the Hydro Venturi, is to be tested in San Francisco Bay.

In late April 2008, Ocean Renewable Power Company, LLC (ORPC) successfully completed the testing of its proprietary turbine-generator unit (TGU) prototype at ORPC's Cobscook Bay and Western Passage tidal sites near Eastport, Maine. The TGU is the core of the OCGen technology and utilizes advanced design cross-flow (ADCF) turbines to drive a permanent magnet generator located between the turbines and mounted on the same shaft. ORPC has developed TGU designs that can be used for generating power from river, tidal and deep water ocean currents.

Trials in the Strait of Messina, Italy, started in 2001 of the Kobold concept.

Commercial Plans

RWE's npower announced that it is in partnership with Marine Current Turbines to build a tidal farm of SeaGen turbines off the coast of Anglesey in Wales.

In November 2007, British company Lunar Energy announced that, in conjunction with E.ON, they would be building the world's first tidal energy farm off the coast of Pembrokshire in Wales. It will be the world's first deep-sea tidal-energy farm and will provide electricity for 5,000 homes. Eight underwater turbines, each 25 metres long and 15 metres high, are to be installed on the sea bottom off St David's peninsula. Construction is due to start in the summer of 2008 and the proposed tidal energy turbines, described as "a wind farm under the sea", should be operational by 2010.

British Columbia Tidal Energy Corp. plans to deploy at least three 1.2 MW turbines in the Campbell River or in the surrounding coastline of British Columbia by 2009.

An organisation named *Alderney Renewable Energy Ltd* is planning to use tidal turbines to extract power from the notoriously strong tidal races around Alderney in the Channel Islands. It is estimated that up to 3GW could be extracted. This would not only supply the island's needs but also leave a considerable surplus for export.

Nova Scotia Power has selected OpenHydro's turbine for a tidal energy demonstration project in the Bay of Fundy, Nova Scotia, Canada and Alderney Renewable Energy Ltd for the supply of tidal turbines in the Channel Islands. Open Hydro

Energy Calculations

Various turbine designs have varying efficiencies and therefore varying power output. If the efficiency of the turbine "Cp" is known the equation below can be used to determine the power output. The energy available from these kinetic systems can be expressed as:

$$P = Cp \times 0.5 \times \rho \times A \times V^3$$

where:

Cp is the turbine coefficient of performance

P = the power generated (in watts)

ρ = the density of the water (seawater is 1025 kg/m^3)

A = the sweep area of the turbine (in m^2)

V^3 = the velocity of the flow cubed (i.e. V x V x V)

Relative to an open turbine in free stream, shrouded turbines are capable of efficiencies as much as 3 to 4 times the power of the same turbine in open flow.

Potential Sites

As with wind power, selection of location is critical for the tidal turbine. Tidal stream systems need to be located in areas with fast currents where natural flows are concentrated between obstructions, for example at the entrances to bays and rivers, around rocky points, headlands, or between islands or other land masses. The following potential sites have been suggested:

- Cook Inlet in Alaska
- Pentland Firth in Scotland
- Dee estuary in Wales
- Pembrokeshire in Wales
- River Severn between Wales and England
- Solway estuary (Morecambe Bay) in England
- Humber estuary in England
- Mersey river in England
- Channel Islands in the English Channel, off the French coast
- Cook Strait in New Zealand
- Strait of Gibraltar
- Bosporus in Turkey
- Bass Strait in Australia
- Torres Strait in Australia
- Strait of Malacca between Indonesia and Singapore
- Bay of Fundy in Canada.
- East River in New York City
- Vancouver Island in Canada
- Strait of Magellan south of mainland Chile

- Golden Gate in the San Francisco Bay
- Piscataqua River in New Hampshire
- The Race of Alderney and The Swinge in the Channel Islands
- Kalpsar in Gujarat, INDIA

BARRAGE TIDAL POWER

With only a few operating plants globally (a large 240 MW plant on the Rance River, and two small plants, one on the Bay of Fundy and the other across a tiny inlet in Kislaya Guba Russia), the barrage method of extracting tidal energy involves building a barrage across a bay or river as in the case of the Rance tidal power plant in France. Turbines installed in the barrage wall generate power as water flows in and out of the estuary basin, bay, or river. These systems are similar to a hydro dam that produces Static Head or pressure head (a height of water pressure). When the water level outside of the basin or lagoon changes relative to the water level inside, the turbines are able to produce power. The largest such installation has been working on the Rance river, France, since 1966 with an installed (peak) power of 240 MW, and an annual production of 600 GWh (about 68 MW average power). The basic elements of a barrage are caissons, embankments, sluices, turbines, and ship locks. Sluices, turbines, and ship locks are housed in caissons (very large concrete blocks). Embankments seal a basin where it is not sealed by caissons. The sluice gates applicable to tidal power are the flap gate, vertical rising gate, radial gate, and rising sector. Barrage systems are affected by problems of high civil infrastructure costs associated with what is in effect a dam being placed across estuarine systems, and the environmental problems associated with changing a large ecosystem.

Ebb Generation

The basin is filled through the sluices until high tide. Then the sluice gates are closed. (At this stage there may be "Pumping" to raise the level further). The turbine gates are kept closed until the sea level falls to create sufficient head across the barrage, and then are opened so that the turbines generate until the head is again low. Then the sluices are opened, turbines disconnected and the basin is filled again. The cycle repeats itself. Ebb generation (also known as outflow generation) takes its name because generation occurs as the tide changes tidal direction.

Flood Generation

The basin is filled through the turbines, which generate at tide flood. This is generally much less efficient than ebb generation, because the volume contained in the upper half of the basin (which is where ebb generation operates) is greater than the volume of the lower half (filled first during flood generation). Therefore the available level difference — important for the turbine power produced — between the basin side and the sea side of the barrage, reduces more quickly than it would in ebb generation. Rivers flowing into the basin may further reduce the energy potential, instead of enhancing it as in ebb generation. Which of course is not a problem with the "lagoon" model, without river inflow.

Pumping

Turbines are able to be powered in reverse by excess energy in the grid to increase the water level in the basin at high tide (for ebb generation). This energy is more than returned during generation, because power output is strongly related to the head. If water is raised 2 ft (61 cm) by pumping on a high tide of 10 ft (3 m), this will have been raised by 12 ft (3.7 m) at low tide. The cost of a 2 ft rise is returned by the benefits of a 12 ft rise. This is since the correlation between the potential energy is not a linear relationship, rather, is related by the square of the tidal height variation.

Two-basin Schemes

Another form of energy barrage configuration is that of the dual basin type. With two basins, one is filled at high tide and the other is emptied at low tide. Turbines are placed between the basins. Two-basin schemes offer advantages over normal schemes in that generation time can be adjusted with high flexibility and it is also possible to generate almost continuously.

In normal estuarine situations, however, two-basin schemes are very expensive to construct due to the cost of the extra length of barrage. There are some favourable geographies, however, which are well suited to this type of scheme.

Environmental Impact

The placement of a barrage into an estuary has a considerable effect on the water inside the basin and on the ecosystem. Many governments have been reluctant in recent times to grant approval for tidal barrages. Through research conducted on tidal plants, it has been found that tidal barrages constructed at the mouths of estuaries pose similar environmental threats as large dams. The construction of large tidal plants alters the flow of saltwater in and out of estuaries, which changes the hydrology and salinity and possibly negatively affects the marine mammals that use the estuaries as their habitat The La Rance plant, off the Brittany coast of northern France, was the first and largest tidal barrage plant in the world. It is also the only site where a full-scale evaluation of the ecological impact of a tidal power system, operating for 20 years, has been made. French researchers found that the isolation of the estuary during the construction phases of the tidal barrage was detrimental to flora and fauna, however; after ten years, there has been a "variable degree of biological adjustment to the new environmental conditions."

Some species lost their habitat due to La Rance's construction, but other species colonized the abandoned space, which caused a shift in diversity. Also as a result of the construction, sandbanks disappeared, the beach of St. Servan was badly damaged and high-speed currents have developed near sluices, which are water channels controlled by gates.

Turbidity

Turbidity (the amount of matter in suspension in the water) decreases as a result of smaller volume of water being exchanged between the basin and the sea. This lets light from the Sun to penetrate the water further, improving conditions for the phytoplankton. The changes propagate up the food chain, causing a general change in the ecosystem.

TIDAL FENCES AND TURBINES

Tidal fences and turbines can have varying environmental impacts depending on whether or not fences and turbines are constructed with regard to the environment. The main environmental impact of turbines is their impact on fish. If the turbines are moving slowly enough, such as low velocities of 25-50 rpm, fish kill is minimalized and silt and other nutrients are able to flow through the structures For example, a 20kW tidal turbine prototype built in the St. Lawrence Seaway in 1983 reported no fish kills. Tidal fences block off channels, which makes it difficult for fish and wildlife to migrate through those channels.

In order to reduce fish kill, fences could be engineered so that the spaces between the caisson wall and the rotor foil are large enough to allow fish to pass through. Larger marine mammals such as seals or dolphins can be protected from the turbines by fences or a sonar sensor auto-breaking system that automatically shuts the turbines down when marine mammals are detected.

Overall, many researches have argued that while tidal barrages pose environmental threats, tidal fences and tidal turbines, if constructed properly, are likely to be more environmentally benign. Unlike barrages, tidal fences and turbines do not block channels or estuarine mouths, interrupt fish migration or alter hydrology, thus, these options offer energy generating capacity without dire environmental impacts.

Salinity

As a result of less water exchange with the sea, the average salinity inside the basin decreases, also affecting the ecosystem. "Tidal Lagoons" do not suffer from this problem.

Sediment Movements

Estuaries often have high volume of sediments moving through them, from the rivers to the sea. The introduction of a barrage into an estuary may result in sediment accumulation within the barrage, affecting the ecosystem and also the operation of the barrage.

Fish

Fish may move through sluices safely, but when these are closed, fish will seek out turbines and attempt to swim through them. Also, some fish will be unable to escape the water speed near a turbine and will be sucked through. Even with the most fish-friendly turbine design, fish mortality per pass is approximately 15% (from pressure drop, contact with blades, cavitation, etc.). Alternative passage technologies (fish ladders, fish lifts, etc.) have so far failed to solve this problem for tidal barrages, either offering extremely expensive solutions, or ones which are used by a small fraction of fish only. Research in sonic guidance of fish is ongoing. The Open-Centre turbine reduces this problem allowing fish to pass through the open centre of the turbine. Recently a run of the river type turbine has been developed in France. This is a very large slow rotating Kaplan type turbine mounted on an angle. Testing for fish mortality has indicated fish mortality figures to be less than 5%. This concept also seems very suitable for adaption to marine current/tidal turbines.

Energy Calculations

The energy available from barrage is dependent on the volume of water. The potential energy contained in a volume of water is:

$$E = \frac{1}{2} A \rho g h^2$$

where:

- h is the vertical tidal range,
- A is the horizontal area of the barrage basin,
- ρ is the density of water = 1025 kg per cubic meter (seawater varies between 1021 and 1030 kg per cubic meter) and
- g is the acceleration due to the Earth's gravity = 9.81 meters per second squared.

The factor half is due to the fact, that as the basin flows empty through the turbines, the hydraulic head over the dam reduces. The maximum head is only available at the moment of low water, assuming the high water level is still present in the basin.

EXAMPLE CALCULATION OF TIDAL POWER GENERATION

Assumptions:

- Let us assume that the tidal range of tide at a particular place is 32 feet = 10 m (approx)
- The surface of the tidal energy harnessing plant is $9\ km^2$ (3 km × 3 km)= 3000 m × 3000 m = $9 \times 10^6\ m^2$
- Specific density of sea water = 1025.18 kg/m^3

Mass of the water = volume of water × specific gravity

= (area × tidal range) of water × mass density

= $(9 \times 10^6\ m^2 \times 10\ m) \times 1025.18\ kg/m^3$

= 92×10^9 kg (approx)

Potential energy content of the water in the basin at high tide = ½ × area × density × gravitational acceleration × tidal range squared

= $\frac{1}{2} \times 9 \times 10^6\ m^2 \times 1025\ kg/m^3 \times 9.81\ m/s^2 \times (10\ m)^2$

= 4.5×10^{12} J (approx)

Now we have 2 high tides and 2 low tides every day. At low tide the potential energy is zero.

Therefore the total energy potential per day = Energy for a single high tide × 2 = 4.5×10^{12} J × 2 = 9×10^{12} J

Therefore, the mean power generation potential = Energy generation potential / time in 1 day

= 9×10^{12} J / 86400 s = 104 MW

Assuming the power conversion efficiency to be 30%: The daily-average power generated = 104 MW * 30% / 100%

= 31 MW (approx)

A barrage is best placed in a location with very high-amplitude tides. Suitable locations are found in Russia, USA, Canada, Australia, Korea, the UK. Amplitudes of up to 17 m (56 ft) occur for example in the Bay of Fundy, where tidal resonance amplifies the tidal range.

Economics

Tidal barrage power schemes have a high capital cost and a very low running cost. As a result, a tidal power scheme may not produce returns for many years, and investors may be reluctant to participate in such projects. Governments may be able to finance tidal barrage power, but many are unwilling to do so also due to the lag time before investment return and the high irreversible commitment. For example the energy policy of the United Kingdom recognizes the role of tidal energy and expresses the need for local councils to understand the broader national goals of renewable energy in approving tidal projects. The UK government itself appreciates the technical viability and siting options available, but has failed to provide meaningful incentives to move these goals forward.

MATHEMATICAL MODELLING OF TIDAL SCHEMES

In mathematical modelling of a scheme design, the basin is broken into segments, each maintaining its own set of variables. Time is advanced in steps. Every step, neighbouring segments influence each other and variables are updated.

The simplest type of model is the *flat estuary* model, in which the whole basin is represented by one segment. The surface of the basin is assumed to be flat, hence the name. This model gives rough results and is used to compare many designs at the start of the design process.

In these models, the basin is broken into large segments (1D), squares (2D) or cubes (3D). The complexity and accuracy increases with dimension.

Mathematical modelling produces quantitative information for a range of parameters, including:

- Water levels (during operation, construction, extreme conditions, etc.)
- Currents
- Waves
- Power output
- Turbidity

- Salinity
- Sediment movements

GLOBAL ENVIRONMENTAL IMPACT

A tidal power scheme is a long-term source of electricity. A proposal for the Severn Barrage, if built, has been projected to save 18 million tonnes of coal per year of operation. This decreases the output of greenhouse gases into the atmosphere. If fossil fuel resources decline during the 21st century, as predicted by Hubbert peak theory, tidal power is one of the alternative sources of energy that will need to be developed to satisfy the human demand for energy.

Operating Tidal Power Schemes

- The first tidal power station was the Rance tidal power plant built over a period of 6 years from 1960 to 1966 at La Rance, France. It has 240 MW installed capacity.
- The first tidal power site in North America is the Annapolis Royal Generating Station, Annapolis Royal, Nova Scotia, which opened in 1984 on an inlet of the Bay of Fundy. It has 18 MW installed capacity.
- The first in-stream tidal current generator in North America (Race Rocks Tidal Power Demonstration Project) was installed at [(Race Rocks]] on southern Vancouver Island in September 2006. The next phase in the development of this tidal current generator will be in Nova Scotia.
- A small project was built by the Soviet Union at Kislaya Guba on the Barents Sea. It has 0.5 MW installed capacity. In 2006 it was upgraded with 1.2MW experimental advanced orthogonal turbine.
- 1.2 MW SeaGen system became operational in late 2008 on Strangford Lough in Northern Ireland.

TIDAL POWER SCHEMES BEING CONSIDERED

In the table, "-" indicates missing information, "?" indicates information which has not been decided

Country	Place	Mean tidal range (m)	Area of basin (km²)	Maximum capacity (MW)
Argentina	San Jose	5.9	-	6800
Australia	Secure Bay	10.9	-	?
Canada	Cobequid	12.4	240	5338
	Cumberland	10.9	90	1400
	Shepody	10.0	115	1800
	Passamaquoddy	5.5	-	?
India	Kutch	5.3	170	900
	Cambay	6.8	1970	7000
South Korea	Garolim	4.7	100	480
	Cheonsu	4.5	-	-
Mexico	Rio Colorado	6-7	-	?
	Tiburon	-	-	?
United Kingdom and Channel Islands	River Severn	7.8	450	8640
	River Mersey	6.5	61	700
	Conwy	5.2	5.5	33
	Alderney	6.9	N/A	3000
United States	Passamaquoddy Bay, Maine	5.5	-	?
	Knik Arm, Alaska	7.5	-	2900
	Turnagain Arm, Alaska	7.5	-	6501
	Golden Gate, California	?	-	?
Russia	Mezen	9.1	2300	19200
	Tugur	-	-	8000
	Penzhinskaya Bay	6.0	20,500	87,000
South Africa	Mozambique Channel	?	?	?
New Zealand	Kaipara Harbour	2.10	947	200
	Cook Strait	-	n/a	10

- A 12MW project at Kislaya Guba in Russia with orthogonal turbines is under construction.
- China has apparently developed several small tidal power projects and one large facility in Jiangxia.

- China is developing a tidal lagoon near the mouth of the Yalu.
- Scotland has committed to having 18% of its power from green sources by 2010, including 10% from a tidal generator. The British government says this will replace one huge fossil fuelled power station.
- South African energy parastatal Eskom is investigating using the Mozambique Current to generate power off the coast of KwaZulu Natal. Because the continental shelf is near to land it may be possible to generate electricity by tapping into the fast flowing Mozambique current.

4

Wind Power

Wind power is the conversion of wind energy into a useful form, such as electricity, using wind turbines. At the end of 2008, worldwide nameplate capacity of wind-powered generators was 121.2 gigawatts. Although wind produces only about 1.5% of worldwide electricity use, it is growing rapidly, having doubled in the three years between 2005 and 2008. In several countries it has achieved relatively high levels of penetration, accounting for approximately 19% of electricity production in Denmark, 11% in Spain and Portugal, and 7% in Germany and the Republic of Ireland in 2008.

Wind energy has historically been used directly to propel sailing ships or converted into mechanical energy for pumping water or grinding grain, but the principal application of wind power today is the generation of electricity. Wind power, along with solar power, is non-dispatchable, meaning that for economic operation all of the available output must be taken when it is available, and other resources, such as hydroelectricity, must be used to match supply with demand.

Large scale wind farms are typically connected to the local electric power transmission network, with smaller turbines being used to provide electricity to isolated locations. Utility companies increasingly buy back surplus electricity produced by small domestic turbines. Wind energy as a power source is favoured by many environmentalists as an alternative to fossil fuels, as it is plentiful, renewable, widely distributed, clean, and produces lower greenhouse gas emissions, although the construction of wind farms is not universally welcomed due to their visual impact and other effects on the environment. The intermittency of wind seldom creates problems when using wind power to supply a low proportion of total demand. Where wind is to be used for a moderate fraction of demand, additional costs for compensation of intermittency are considered to be modest.

Although the wind power industry will be impacted by the global financial crisis in 2009 and 2010, a BTM Consult five year forecast up to 2013 projects substantial growth. Over the past five years the average growth in new installations has been 27.6 per cent each year. In the forecast to 2013 the expected average annual growth rate is 15.7 per cent. More than 200 GW of new wind power capacity could come on line before the end of 2013. Wind power market penetration is expected to reach 3.35 per cent by 2013 and 8 per cent by 2018.

HISTORY

Humans have been using wind power for at least 5,500 years to propel sailboats and sailing ships, and architects have used wind-driven natural ventilation in buildings since similarly ancient times. The use of wind to provide mechanical power came somewhat later in antiquity.

The Babylonian emperor Hammurabi planned to use wind power for his ambitious irrigation project in the 17th century BC. The ancient Sinhalese utilized the monsoon winds to power furnaces as early as 300 BC evidence has been found in cities such as Anuradhapura and in other cities around Sri Lanka The furnaces were constructed on the path of the monsoon winds to exploit the wind power, to bring the temperatures inside up to 1100-1200 Celsius. An early historical reference to a rudimentary windmill was used to power an organ in the 1st century AD. The first practical windmills were later built in Sistan, Afghanistan, from the 7th century. These were vertical-axle windmills, which had long vertical driveshafts with rectangle shaped blades. Made of six to twelve sails covered in reed matting or cloth material, these windmills were used to grind corn and draw up water, and were used in the gristmilling and sugarcane industries. Horizontal-axle windmills were later used extensively in Northwestern Europe to grind flour beginning in the 1180s, and many Dutch windmills still exist.

In the United States, the development of the "water-pumping windmill" was the major factor in allowing the farming and ranching of vast areas of North America, which were otherwise devoid of readily accessible water. They contributed to the expansion of rail transport systems throughout the world, by pumping water from wells for the steam locomotives. The multi-bladed wind turbine atop a lattice tower made of wood or steel was, for many years, a fixture of the landscape throughout rural America.

The first modern wind turbines were built in the early 1980s, although more efficient designs are still being developed.

WIND ENERGY

Distribution of wind speed (red) and energy (blue) for all of 2002 at the Lee Ranch facility in Colorado. The histogram shows measured data, while the curve is the Rayleigh model distribution for the same average wind speed. Energy is the Betz limit through a 100 meter diameter circle facing directly into the wind. Total energy for the year through that circle was 15.4 gigawatt-hours.

The Earth is unevenly heated by the sun resulting in the poles receiving less energy from the sun than the equator does. Also, the dry land heats up (and cools down) more quickly than the seas do. The differential heating drives a global atmospheric convection system reaching from the Earth's surface to the stratosphere which acts as a virtual ceiling. Most of the energy stored in these wind movements can be found at high altitudes where continuous wind speeds of over 160 km/h (100 mph) occur. Eventually, the wind energy is converted through friction into diffuse heat throughout the Earth's surface and the atmosphere.

The total amount of economically extractable power available from the wind is considerably more than present human power use from all sources. An estimated 72 TW of wind power on the Earth potentially can be commercially viable, compared to about 15 TW average global power consumption from all sources in 2005. Not all the energy of the wind flowing past a given point can be recovered.

Distribution of Wind Speed

The strength of wind varies, and an average value for a given location does not alone indicate the amount of energy a wind turbine could produce there. To assess the frequency of wind speeds at a particular location, a probability distribution function is often fit to the observed data. Different locations will have different wind speed distributions. The Rayleigh model closely mirrors the actual distribution of hourly wind speeds at many locations. Because so much power is generated by higher wind speed, much of the energy comes in short bursts. The 2002 Lee Ranch sample is telling; half of the energy available arrived in just 15% of the operating time. The consequence is that wind energy from a particular turbine or wind farm does not have as consistent an output as fuel-fired power plants; utilities that use wind power provide power from starting existing generation for times when the wind is weak thus wind power is primarily a fuel saver rather than a capacity saver. Making wind power more consistent requires that various existing technologies and methods be extended, in particular the use of stronger inter-regional transmission to link widely distributed wind farms, since the average variability is much less; the use of hydro storage and demand-side energy management.

Wind power density (WPD) is a calculation relating to the effective force of the wind at a particular location, frequently expressed in terms of the elevation above ground level over a period of time. It further takes into account wind velocity and mass. Color coded maps are frequently prepared for a particular area, described for example as "Mean Annual Power Density at 70 Meters." The results of the above calculation are used in an index developed by the National Renewable Energy Labs and referred to as "NREL CLASS." The larger the WPD calculation, the higher it is rated by class. Even though wind power is comparable in Texas and Kansas, there are about 10 times as many wind turbines in Texas as there are in Kansas.

ELECTRICITY GENERATION

Grid Management System

Electricity generated by a wind farm is normally fed into the national electric power transmission network. Individual turbines are interconnected with a medium voltage (usually 34.5 kV) power collection system and communications network. At a substation, this medium-voltage electrical current is increased in voltage with a transformer for connection to the high voltage transmission system. The surplus power produced by domestic microgenerators can, in some jurisdictions, be fed back into the network and sold back to the utility company, producing a retail credit for the consumer to offset their energy costs.

Induction generators, often used for wind power projects, require reactive power for excitation so substations used in wind-power collection systems include substantial capacitor banks for power factor correction. Different types of wind turbine generators behave differently during transmission grid disturbances, so extensive modelling of the dynamic electromechanical characteristics of a new wind farm is required by transmission system operators to ensure predictable stable behaviour during system faults. In particular, induction generators cannot support the system voltage during faults, unlike steam or hydro turbine-driven synchroncus generators (however, properly matched power factor correction capacitors along with electronic control of resonance can support induction generation without grid). Doubly-fed machines, or wind turbines with solid-state converters between the turbine generator and the collector system, have generally more desirable properties for grid interconnection. Transmission systems operators will supply a wind farm developer with a *grid code* to specify the requirements for interconnection to the transmission grid. This will include power factor, constancy of frequency and dynamic behaviour of the wind farm turbines during a system fault.

Capacity Factor

Since wind speed is not constant, a wind farm's annual energy production is never as much as the sum of the generator nameplate ratings multiplied by the total hours in a year. The ratio of actual productivity in a year to this theoretical maximum is called the capacity factor. Typical capacity factors are 20-40%, with values at the upper end of the range in particularly favourable sites. For example, a 1 megawatt turbine with a capacity factor of 35% will not produce 8,760 megawatt-hours in a year (1x24x365), but only 1x0.35x24x365 = 3,066 MWh, averaging to 0.35 MW. Online data is available for some locations and the capacity factor can be calculated from the yearly output.

Unlike fueled generating plants, the capacity factor is limited by the inherent properties of wind. Capacity factors of other types of power plant are based mostly on fuel cost, with a small amount of downtime for maintenance. Nuclear plants have low incremental fuel cost, and so are run at full output and achieve a 90% capacity factor. Plants with higher fuel cost are throttled back to follow load. Gas turbine plants using natural gas as fuel may be very expensive to operate and may be run only to meet peak power demand. A gas turbine plant may have an annual capacity factor of 5-25% due to relatively high energy production cost.

According to a 2007 Stanford University study published in the Journal of Applied Meteorology and Climatology, interconnecting ten or more wind farms can allow an average of 33% of the total energy produced to be used as reliable, baseload electric power, as long as minimum criteria are met for wind speed and turbine height.

INTERMITTENCY AND PENETRATION LIMITS

Electricity generated from wind power can be highly variable at several different timescales: from hour to hour, daily, and seasonally. Annual variation also exists, but is not as significant. Because instantaneous electrical generation and consumption must remain in balance to maintain grid stability, this variability can present substantial challenges to incorporating large amounts of wind power into a grid system. Intermittency and the non-dispatchable nature of wind energy production can raise costs for regulation, incremental operating reserve, and (at high penetration levels) could require an increase in the already existing energy demand management, load shedding, or storage solutions or system interconnection with HVDC cables. At low levels of wind penetration, fluctuations in load and allowance for failure of large generating units requires reserve capacity that can also regulate for variability of wind generation.

A series of detailed modelling studies which looked at the Europe wide adoption of renewable energy and interlinking power grids using HVDC cables, indicates that the entire power usage could come from renewables, with 70% total energy from wind at the same sort of costs or lower than at present. Intermittency would be dealt with, according to this model, by a combination of geographic dispersion to de-link weather system effects, and the ability of HVDC to shift power from windy areas to non-windy areas.

Pumped-storage hydroelectricity or other forms of grid energy storage can store energy developed by high-wind periods and release it when needed. Stored energy increases the economic value of wind energy since it can be shifted to displace higher cost generation during peak demand periods. The potential revenue from this arbitrage can offset the cost and losses of storage; the cost of storage may add 25% to the cost of any wind energy stored, but it is not envisaged that this would apply to a large proportion of wind energy generated. Thus the 2 GW Dinorwig pumped storage plant adds costs to nuclear energy in the UK for which it was built, but not to all the power produced from the 30 or so GW of nuclear plants in the UK.

In particular geographic regions, peak wind speeds may not coincide with peak demand for electrical power. In California and Texas, for example, hot days in summer may have low wind speed and high electrical demand due to air conditioning. Some utilities subsidize the purchase of geothermal heat pumps by their customers, to reduce electricity demand during the summer months by making air conditioning up to 70% more efficient; widespread adoption of this technology would better match electricity demand to wind availability in areas with hot summers and low summer winds. Geothermal heat pumps also allow renewable electricity from wind to displace natural gas and heating oil for central heating during winter, when winds tend to be stronger in many areas. Another option is to interconnect widely dispersed geographic areas with a relatively cheap and efficient HVDC "Super grid". In the USA it is estimated that to upgrade the transmission system to take in planned or potential renewables would cost at least $60 billion. Total annual US power consumption in 2006 was 4 thousand billion kilowatt hours. Over an asset life of 40 years and low cost utility investment grade funding, the cost of $60 billion investment would be about 5% p.a. ie $3 billion p.a. Dividing by total power used gives an increased unit cost of around $3,000,000,000 x 100 / 4,000 x 1 exp9 = 0.075 cent / kWh.

According to a 2007 Stanford University study published in the Journal of Applied Meteorology and Climatology, interconnecting ten or more wind farms allows 33 to 47% of the total energy produced to be used as reliable, baseload electric power, as long as minimum criteria are met for wind speed and turbine height.

In the UK, demand for electricity is higher in winter than in summer, and so are wind speeds. Solar power tends to be complementary to wind. On daily to weekly timescales, high pressure areas tend to bring clear skies and low surface winds, whereas low pressure areas tend to be windier and cloudier. On seasonal timescales, solar energy typically peaks in summer, whereas in many areas wind energy is lower in summer and higher in winter. Thus the intermittencies of wind and solar power tend to cancel each other somewhat. A demonstration project at the Massachusetts Maritime Academy shows the effect. The Institute for Solar Energy Supply Technology of the University of Kassel pilot-tested a combined power plant linking solar, wind, biogas and hydrostorage to provide load-following power around the clock, entirely from renewable sources.

A report from Denmark noted that their wind power network was without power for 54 days during 2002. Wind power advocates argue that these periods of low wind can be dealt with by simply restarting existing power stations that have been held in readiness or interlinking with HVDC. The cost of keeping a fossil fuel power station idle is in fact quite low, since the main cost of running a power station is the fuel (see spark spread and dark spread).

Penetration

Wind energy "penetration" refers to the fraction of energy produced by wind compared with the total available generation capacity. There is no generally accepted "maximum" level of wind penetration. The limit for a particular grid will depend on the existing generating plants, pricing mechanisms, capacity for storage or demand management, and other factors. An interconnected electricity grid will already include reserve generating and transmission capacity to allow for equipment failures; this reserve capacity can also serve to regulate for the varying power generation by wind plants. Studies have indicated that 20% of the total electrical energy consumption may be incorporated with minimal difficulty. These studies have been for locations with geographically dispersed wind farms, some degree of dispatchable energy, or hydropower with storage capacity, demand management, and interconnection to a large grid area export of electricity when needed. Beyond this level, there are few technical limits, but the economic implications become more significant.

However In evidence to the House of Lords Economic Affairs Select Committee, the UK System Operator, National Grid have quoted estimates of balancing costs for 40% wind and these lie in the range £500-1000M per annum. "These balancing costs represent an additional £6 to £12 per annum on average consumer electricity bill of around £390."

At present, few grid systems have penetration of wind energy above 5%: Denmark (values over 19%), Spain and Portugal (values over 11%), Germany and the Republic of Ireland (values over 6%). The Danish grid is heavily interconnected to the European electrical grid, and it has solved grid management problems by exporting almost half of its wind power to Norway. The correlation between electricity export and wind power production is very strong. Denmark has active plans to increase the percentage of power generated to over 50%.

A study commissioned by the state of Minnesota considered penetration of up to 25%, and concluded that integration issues would be manageable and have incremental costs of less than one-half cent ($0.0045) per kWh.

ESB National Grid, Ireland's electric utility, in a 2004 study that, concluded that to meet the renewable energy targets set by the EU in 2001 would "increase electricity generation costs by a modest 15%." A recent report by Sinclair Merz saw no difficulty in accommodating 50% of total power delivered in the UK at modest cost increases.

Predictability

Related to variability is the short-term (hourly or daily) predictability of wind plant output. Like other electricity sources, wind energy must be "scheduled". The nature of this energy source makes it inherently variable. Wind power forecasting methods are used, but predictability of wind plant output remains low for short-term operation.

Turbine Placement

Good selection of a wind turbine site is critical to economic development of wind power. Aside from the availability of wind itself, other factors include the availability of transmission lines, value of energy to be produced, cost of land acquisition, land use considerations, and environmental impact of construction and operations. Off-shore locations may offset their higher construction cost with higher annual load factors, thereby reducing cost of energy produced. Wind farm designers use specialized wind energy software applications to evaluate the impact of these issues on a given wind farm design.

Studies in the UK have shown that if onshore turbines are placed in a straight line then an increased risk of aerodynamic modulation can occur which can result in noise nuisance to nearby residents.

Offshore Wind Farms

As of 2008, Europe leads the world in development of offshore wind power, due to strong wind resources and shallow water in the North Sea and the Baltic Sea, and limitations on suitable locations on land due to dense populations and existing developments. Denmark installed the first offshore wind farms, and for years was the world leader in offshore wind power until the United Kingdom gained the lead in October, 2008 with 590 MW of nameplate capacity installed. The United Kingdom planned to build much more extensive offshore wind farms by 2020. Other large markets for wind power, including the United States and China focused first on developing their on-land wind resources where construction costs are lower (such as in the Great Plains of the U.S., and the similarly wind-swept steppes of Xinjiang and Inner Mongolia in China), but population centers along coastlines in many parts of the world are close to offshore wind resources, which would reduce transmission costs.

On 21 December 2007, Q7 (later renamed as Princess Amalia Wind Farm) exported first power to the Dutch grid, which was a milestone for the offshore wind industry. The 120MW offshore wind farm with a construction budget of □383 million was the first to be financed by a nonrecourse loan (project finance). The project comprises 60 Vestas V80-2MW wind turbines. Each turbine's tower rests on a monopile foundation to a depth of between 18-23 meters at a distance of about 23 km off the Dutch coast.

Transporting large wind turbine components (tower sections, nacelles, and blades) is much easier over water than on land, because ships and barges can handle large loads more easily than trucks/lorries or trains. On land, large goods vehicles must negotiate bends on roadways, which fixes the maximum length of a wind turbine blade that can move from point to point on the road network; no such limitation exists for transport on open water. Construction and maintenance costs per wind turbine are higher for offshore wind farms, motivating operators to reduce the number of wind turbines for a given total power by installing the largest available units. An example is Belgium's Thorntonbank Wind Farm with construction underway in 2008, featuring 5MW wind turbines from REpower, which were among the largest wind turbines in the world at the time.

UTILIZATION OF WIND POWER

Installed windpower capacity (MW)

#	Nation	2005	2006	2007	2008
1	United States	9,149	11,603	16,818	25,170
2	Germany	18,415	20,622	22,247	23,903
3	Spain	10,028	11,615	15,145	16,740
4	China	1,260	2,604	6,050	12,210
5	India	4,430	6,270	8,000	9,587
6	Italy	1,718	2,123	2,726	3,736
7	France	757	1,567	2,454	3,404
8	United Kingdom	1,332	1,963	2,389	3,288
9	Denmark (& Faeroe Islands)	3,136	3,140	3,129	3,160
10	Portugal	1,022	1,716	2,150	2,862
11	Canada	683	1,459	1,856	2,369
12	Netherlands	1,219	1,560	1,747	2,225
13	Japan	1,061	1,394	1,538	1,880
14	Australia	708	817	824	1,494
15	Sweden	510	572	788	1,067
16	Ireland	496	745	805	1,245
17	Austria	819	965	982	995
18	Greece	573	746	871	990
19	Poland	83	153	276	472
20	Turkey	20	51	146	333
21	Norway	267	314	333	428
22	Belgium	167	193	287	384
23	Egypt	145	230	310	390
24	Taiwan	104	188	282	358
25	Brazil	29	237	247	338
26	New Zealand	169	171	322	325
27	South Korea	98	173	191	278
28	Bulgaria	6	20	35	158
29	Czech Republic	28	50	116	150
30	Finland	82	86	110	140

31	Morocco	64	124	114	125
32	Hungary	18	61	65	127
33	Ukraine	77	86	89	90
34	Mexico	3	88	87	85
35	Iran	23	48	66	82
36	Costa Rica	71	74	74	74
	Rest of Europe	129	163		
	Rest of Americas	109	109		
	Rest of Asia	38	38		
	Rest of Africa & Middle East	31	31		
	Rest of Oceania	12	12		
	World total (MW)	59,091	74,223	93,849	121,188

The modern wind power industry began in 1979 with the serial production of wind turbines by Danish manufacturers Kuriant, Vestas, Nordtank, and Bonus. These early turbines were small by today's standards, with capacities of 20 to 30 kW each. Since then, they have increased greatly in size, while wind turbine production has expanded to many countries all over the world.

There are now many thousands of wind turbines operating, with a total nameplate capacity of 121,188 MWp of which wind power in Europe accounts for 55% (2008). World wind generation capacity more than quadrupled between 2000 and 2006, doubling about every three years. By comparison, photovoltaics has been doubling about every two years (48%/year), although from a smaller base (15,200 MWp in 2008). 81% of wind power installations are in the US and Europe. The share of the top five countries in terms of new installations fell from 71% in 2004 to 62% in 2006, but climbed to 73% by 2008 as those countries — the United States, Germany, Spain, China, and India — have seen substantial capacity growth in the past two years (see chart).

By 2010, the World Wind Energy Association expects 160GW of capacity to be installed worldwide, up from 73.9 GW at the end of 2006, implying an anticipated net growth rate of more than 21% per year. Denmark generates nearly one-fifth of its electricity with wind turbines — the highest percentage of any country — and is ninth in the world in total wind power generation. Denmark is prominent in the manufacturing and use of wind turbines, with a commitment made in the 1970s to eventually produce half of the country's power by wind.

In recent years, the United States has added more wind energy to its grid than any other country; U.S. wind power capacity grew by 45% to 16.8 gigawatts in 2007 and surpassing Germany's nameplate capacity in 2008. California was one of the incubators of the modern wind power industry, and led the U.S. in installed capacity for many years; however, by the end of 2006, Texas became the leading wind power state and continues to extend its lead. At the end of 2008, the state had 7,116 MW installed, which would have ranked it sixth in the world if Texas was a separate country. Iowa and Minnesota each grew to more than 1 gigawatt installed by the end of 2007; in 2008 they were joined by Oregon, Washington, and Colorado. Wind power generation in the U.S. was up 31.8% in February, 2007 from February, 2006. The average output of one megawatt of wind power is equivalent to the average electricity consumption of about 250 American households. According to the American Wind Energy Association, wind will generate enough electricity in 2008 to power just over 1% (4.5 million households) of total electricity in U.S., up from less than 0.1% in 1999. U.S. Department of Energy studies have concluded wind harvested in the Great Plains states of Texas, Kansas, and North Dakota could provide enough electricity to power the entire nation, and that offshore wind farms could do the same job. In addition, the wind resource over and around the Great Lakes, recoverable with currently available technology, could by itself provide 80% as much power as the U.S. and Canada currently generate from non-renewable resources, with Michigan's share alone equating to one third of current U.S. electricity demand.

India ranks 5th in the world with a total wind power capacity of 9,587 MW in 2008, or 3% of all electricity produced in India. The World Wind Energy Conference in New Delhi in

November 2006 has given additional impetus to the Indian wind industry. Muppandal village in Tamil Nadu state, India, has several wind turbine farms in its vicinity, and is one of the major wind energy harnessing centres in India led by majors like Suzlon, Vestas, Micon among others.

In 2005, China announced it would build a 1000-megawatt wind farm in Hebei for completion in 2020. China has set a generating target of 30,000 MW by 2020 from renewable energy sources — it says indigenous wind power could generate up to 253,000 MW. A Chinese renewable energy law was adopted in November 2004, following the World Wind Energy Conference organized by the Chinese and the World Wind Energy Association. By 2008, wind power was growing faster in China than the government had planned, and indeed faster in percentage terms than in any other large country, having more than doubled each year since 2005. Policymakers doubled their wind power prediction for 2010, after the wind industry reached the original goal of 5 GW three years ahead of schedule. Current trends suggest an actual installed capacity near 20 GW by 2010, with China shortly thereafter pursuing the United States for the world wind power lead.

Mexico recently opened La Venta II wind power project as an important step in reducing Mexico's consumption of fossil fuels. The 88 MW project is the first of its kind in Mexico, and will provide 13 percent of the electricity needs of the state of Oaxaca. By 2012 the project will have a capacity of 3500 MW.

Another growing market is Brazil, with a wind potential of 143 GW. The federal government has created an incentive program, called Proinfa, to build production capacity of 3300 MW of renewable energy for 2008, of which 1422 MW through wind energy. The program seeks to produce 10% of Brazilian electricity through renewable sources.

South Africa has a proposed station situated on the West Coast north of the Olifants River mouth near the town of Koekenaap, east of Vredendal in the Western Cape province. The station is proposed to have a total output of 100MW although there are negotiations to double this capacity. The plant could be operational by 2010.

France has announced a target of 12,500 MW installed by 2010, though their installation trends over the past few years suggest they'll fall well short of their goal.

Canada experienced rapid growth of wind capacity between 2000 and 2006, with total installed capacity increasing from 137 MW to 1,451 MW, and showing an annual growth rate of 38%. Particularly rapid growth was seen in 2006, with total capacity doubling from the 684 MW at end-2005. This growth was fed by measures including installation targets, economic incentives and political support. For example, the Ontario government announced that it will introduce a feed-in tariff for wind power, referred to as 'Standard Offer Contracts', which may boost the wind industry across the province. In Quebec, the provincially-owned electric utility plans to purchase an additional 2000 MW by 2013.

SMALL SCALE WIND POWER

Small scale wind power is the name given to wind generation systems with the capacity to produce 50 kW or less of electrical power. Isolated communities, that otherwise rely on diesel generators, may use wind turbines to displace diesel fuel consumption. Individuals may purchase these systems to reduce or eliminate their dependence on grid electricity for economic or other reasons, or to reduce their carbon footprint. Wind turbines have been used for household electricity generation in conjunction with battery storage over many decades in remote areas. Increasingly, U.S. consumers are choosing to purchase grid-connected turbines in the 1 to 10 kilowatt range to power their whole homes. Household generator units of more than 1 kW are now functioning in several countries, and in every state in the U.S. Grid-connected wind turbines may use grid energy storage, displacing purchased energy with local production when available. Off-grid system users can either adapt to intermittent power or use batteries, photovoltaic or diesel systems to supplement the wind turbine. In urban locations, where it is difficult to obtain predictable or large amounts of wind energy (little is known about the actual wind resource of towns and cities), smaller systems may still be used to run low power equipment. Equipment such as parking meters or wireless Internet gateways may be powered by a wind turbine that charges a small battery, replacing the need for a connection to the power grid, making the potential carbon savings of small wind turbines difficult to determine.

A new Carbon Trust study into the potential of small-scale wind energy has found that small wind turbines could provide up to 1.5 terawatt hours (TW·h) per year of electricity (0.4% of total UK electricity consumption) and 0.6 million tonnes of carbon dioxide (Mt CO_2) emission savings. This is based on 10% of households installing turbines at costs competitive with grid electricity, which is currently around 12p per kilowatt-hour.

Distributed generation from renewable resources is increasing as a consequence of the increased awareness of climate change. The electronic interfaces required to connect renewable generation units with the utility system can include additional functions such as active filtering to enhance the power quality.

ECONOMICS AND FEASIBILITY

Growth and Cost Trends

Wind and hydroelectric power generation have negligible fuel costs and relatively low maintenance costs; in economic terms, wind power has a low marginal cost and a high proportion of capital cost. The estimated average cost per unit incorporates the cost of construction of the turbine and transmission facilities, borrowed funds, return to investors (including cost of risk), estimated annual production, and other components, averaged over the projected useful life of the equipment, which may be in excess of twenty years. Energy cost estimates are highly dependent on these assumptions so published cost figures can differ substantially. A British Wind Energy Association report gives an average generation cost of onshore wind power of around 3.2 cents per kilowatt hour (2005). Cost per unit of energy produced was estimated in 2006 to be comparable to the cost of new generating capacity in the United States for coal and natural gas: wind cost was estimated at $55.80 per MWh, coal at $53.10/MWh and natural gas at $52.50. Other sources in various studies have estimated wind to be more expensive than other sources (see Economics of new nuclear power plants, Clean coal, and Carbon capture and storage).

In 2004, wind energy cost one-fifth of what it did in the 1980s, and some expected that downward trend to continue as larger multi-megawatt turbines were mass-produced. However, installed cost averaged □1,300 per kilowatt in 2007, compared to □1,100 per kilowatt in 2005. Not as many facilities can produce large modern turbines and their towers and foundations, so constraints develop in the supply of turbines resulting in higher costs. Research from a wide variety of sources in various countries shows that support for wind power is consistently between 70 and 80 percent amongst the general public.

Global Wind Energy Council (GWEC) figures show that 2007 recorded an increase of installed capacity of 20 GW, taking the total installed wind energy capacity to 94 GW, up from 74 GW in 2006. Despite constraints facing supply chains for wind turbines, the annual market for wind continued to increase at an estimated rate of 31% following 32% growth in 2006. In terms of economic value, the wind energy sector has become one of the important players in the energy markets, with the total value of new generating equipment installed in 2007 reaching □25 billion, or US$36 billion.

Existing generation capacity represents sunk costs, and the decision to continue production will depend on marginal costs going forward, not estimated average costs at project inception. For example, the estimated cost of new wind power capacity may be lower than that for "new coal" (estimated average costs for new generation capacity) but higher than for "old coal" (marginal cost of production for existing capacity). Therefore, the choice to increase wind capacity will depend on factors including the profile of existing generation capacity.

Theoretical Potential

Wind power available in the atmosphere is much greater than current world energy consumption. The most comprehensive study to date found the potential of wind power on land and near-shore to be 72 TW, equivalent to 54,000 MToE (million tons of oil equivalent) per year, or over five times the world's current energy use in all forms. The potential takes into account only locations with mean annual wind speeds e" 6.9 m/s at 80 m. It assumes 6 turbines per square kilometer for 77 m diameter, 1.5 MW turbines on roughly 13% of the total global land area (though that land would also be available for other compatible uses such as farming). The authors acknowledge that many practical barriers would need to be overcome to reach this theoretical capacity.

The practical limit to exploitation of wind power will be set by economic and environmental factors, since the resource available is far larger than any practical means to develop it.

Direct Costs

Many potential sites for wind farms are far from demand centres, requiring substantially more money to construct new transmission lines and substations. In some regions this is partly because frequent strong winds themselves have discouraged dense human settlement in especially windy areas. The wind which was historically a nuisance is now becoming a valuable rescurce, but it may be far from large populations which developed in areas more sheltered from wind.

Since the primary cost of producing wind energy is construction and there are no fuel costs, the average cost of wind energy per unit of production depends on a few key assumptions, such as the cost of capital and years of assumed service. The marginal cost of wind energy once a plant is constructed is usually less than 1 cent per kilowatt-hour. Since the cost of capital plays a large part in projected cost, risk (as perceived by investors) will affect projected costs per unit of electricity.

The commercial viability of wind power also depends on the pricing regime for power producers. Electricity prices are highly regulated worldwide, and in many locations may not reflect the full cost of production, let alone indirect subsidies or negative externalities. Customers may enter into long-term pricing contracts for wind to reduce the risk of future pricing changes, thereby ensuring more stable returns for projects at the development stage. These may take the form of standard offer contracts, whereby the system operator undertakes to purchase power from wind at a fixed price for a certain period (perhaps up to a limit); these prices may be different than purchase prices from other sources, and even incorporate an implicit subsidy.

In jurisdictions where the price for electricity is based on market mechanisms, revenue for all producers per unit is higher when their production coincides with periods of higher prices. The profitability of wind farms will therefore be higher if their production schedule coincides with these periods. If wind represents a significant portion of supply, average revenue per unit of production may be lower as more expensive and less-efficient forms of generation, which typically set revenue levels, are displaced from economic dispatch. This may be of particular concern if the output of many wind plants in a market have strong temporal correlation. In economic terms, the marginal revenue of the wind sector as penetration increases may diminish.

External Costs

Most forms of energy production create some form of negative externality: costs that are not paid by the producer or consumer of the good. For electric production, the most significant externality is pollution, which imposes social costs in increased health expenses, reduced agricultural productivity, and other problems. In addition, carbon dioxide, a greenhouse gas produced when fossil fuels are burned, may impose even greater costs in the form of global warming. Few mechanisms currently exist to *internalise* these costs, and the total cost is highly uncertain. Other significant externalities can include military expenditures to ensure access to fossil fuels, remediation of polluted sites, destruction of wild habitat, loss of scenery/tourism, etc.

If the external costs are taken into account, wind energy can be competitive in more cases, as costs have generally decreased due to technology development and scale enlargement. Supporters argue that, once external costs and subsidies to other forms of electrical production are accounted for, wind energy is amongst the least costly forms of electrical production. Critics argue that the level of required subsidies, the small amount of energy needs met, the expense of transmission lines to connect the wind farms to population centers, and the uncertain financial returns to wind projects make it inferior to other energy sources. Intermittency and other characteristics of wind energy also have costs that may rise with higher levels of penetration, and may change the cost-benefit ratio.

Incentives

Wind energy in many jurisdictions receives some financial or other support to encourage its development. A key issue is the comparison to other forms of energy production, and their total cost. Two main points of discussion arise: direct subsidies and externalities for various sources of electricity, including wind. Wind energy benefits from subsidies of various kinds in many jurisdictions, either to increase its attractiveness, or to compensate for subsidies received by other forms of production which have significant negative externalities.

In the United States, wind power receives a tax credit for each kilowatt-hour produced; at 1.9 cents per kilowatt-hour in 2006, the credit has a yearly inflationary adjustment. Another tax benefit is accelerated depreciation. Many American states also provide incentives, such as exemption from property tax, mandated purchases, and additional markets for "green credits." Countries such as Canada and Germany also provide incentives for wind turbine construction, such as tax credits or minimum purchase prices for wind generation, with assured grid access (sometimes referred to as feed-in tariffs). These feed-in tariffs are typically set well above average electricity prices. The Energy Improvement and Extension Act of 2008 contains extensions of credits for wind, including microturbines.

Secondary market forces also provide incentives for businesses to use wind-generated power, even if there is a premium price for the electricity. For example, socially responsible manufacturers pay utility companies a premium that goes to subsidize and build new wind power infrastructure. Companies like the Borealis Press print millions of greeting cards every year using this wind-generated power, and in return they can claim that they are making a powerful "green" effort, in addition to using recycled, chlorine-free paper, soy inks, and safe press wash.

Environmental Effects

Wind power consumes no fuel for continuing operation, and has no emissions directly related to electricity production. Operation does not produce carbon dioxide, sulfur dioxide, mercury, particulates, or any other type of air pollution, as fossil fuel power sources do. Wind power plants consume resources in manufacturing and construction. During manufacture of the wind turbine, steel, concrete, aluminum and other materials will have to be made and transported using energy-intensive processes, generally using fossil energy sources.

The initial carbon dioxide emissions "pay back" is claimed by one company to be within about 9 months of operation for their offshore turbines and the British Wind Energy Association claim the average wind farm will pay back the energy used in its manufacture within 3 to 5 months of operation. However, a report to the British House of Lords in 2004 suggested a payback time of 1.1 years, taking into account factors such as plant construction and decommissioning. A shorter period for offshore facilities was given, as the higher capacity factors would more than offset the added energy costs of installation.

Danger to birds is often the main complaint against the installation of a wind turbine. The Center for Biological Diversity filed a lawsuit against wind farm owners for killing tens of thousands of birds at the Altamont Pass Wind Resource Area near San Francisco, California. However, studies show that the number of birds killed by wind turbines is negligible compared to the number that die as a result of other human activities such as traffic, hunting, power lines and high-rise buildings and especially the environmental impacts of using non-clean power sources. For example, in the UK, where there are several hundred turbines, about one bird is killed per turbine per year; 10 million per year are killed by cars alone. The Audubon Society have also come out in support of wind energy generation, claiming that birds are over 10,000 times more likely to be killed by other human-related causes than by a wind turbine.

Migratory bat species appear to be particularly at risk, especially during key movement periods (spring and more importantly in fall). Lasiurines such as the hoary bat, red bat, and the silver-haired bat appear to be most vulnerable at North American sites. Almost nothing is known about current populations of these species and the impact on bat numbers as a result of mortality at windpower locations. Offshore wind sites 10 km or more from shore do not interact with bat populations. Three different environmental organizations had raised objections to a proposed wind farm at Shaffer Mountain in northeastern Somerset County, Pennsylvania, because the wind farm would be a threat to the Indiana bat, which is listed as an endangered species.

The noise created by wind turbines is often cited as an issue, although the noise of large turbines is far less than of smaller turbines.

Aesthetics have also been a concern. The Massachusetts offshore Cape Wind project was delayed for years mainly because of aesthetic concerns.

Other groups have opposed certain proposals for wind farms on environmental grounds. Environmentalists filed lawsuits to block proposed wind farms in southern Texas, expressing concerns over wetlands, habitat, endangered species and migratory birds, in western Maryland in offshore Hawaii and in New Zealand.

5

Anaerobic Digestion

Anaerobic digestion is a series of processes in which microorganisms break down biodegradable material in the absence of oxygen. It is widely used to treat wastewater sludges and organic waste because it provides volume and mass reduction of the input material. As part of an integrated waste management system, anaerobic digestion reduces the emission of landfill gas into the atmosphere. Anaerobic digestion is a renewable energy source because the process produces a methane and carbon dioxide rich biogas suitable for energy production helping replace fossil fuels. Also, the nutrient-rich solids left after digestion can be used as fertiliser.

The digestion process begins with bacterial hydrolysis of the input materials in order to break down insoluble organic polymers such as carbohydrates and make them available for other bacteria. Acidogenic bacteria then convert the sugars and amino acids into carbon dioxide, hydrogen, ammonia, and organic acids. Acetogenic bacteria then convert these resulting organic acids into acetic acid, along with additional ammonia, hydrogen, and carbon dioxide. Methanogens, finally are able to convert these products to methane and carbon dioxide.

Previously, the technical expertise required to maintain anaerobic digesters coupled with high capital costs and low process efficiencies had limited the level of its industrial application as a waste treatment technology. Anaerobic digestion facilities have, however, been recognised by the United Nations Development Programme as one of the most useful decentralised sources of energy supply, as they are less capital intensive than large power plants.

History

Scientific interest in the gasses produced by the natural decomposition of organic matter, was first reported in the seventeenth century by Robert Boyle and Stephen Hale, who noted that flammable gas was released by disturbing the sediment of streams and lakes. In 1808, Sir Humphry Davy determined that methane was present in the gasses produced by cattle manure. The first anaerobic digester was built by a leper colony in Bombay, India in 1859. In 1895 the technology was developed in Exeter, England, where a septic tank was used to generate gas for street lighting. Also in England, in 1904, the first dual purpose tank for both sedimentation and sludge treatment was installed in Hampton. In 1907, in Germany, a patent was issued for the Imhoff tank, an early form of digester.

Through scientific research anaerobic digestion gained academic recognition in the 1930s. This research led to the discovery of anaerobic bacteria, the microorganisms that facilitate the process. Further research was carried out to investigate the conditions under which methanogenic bacteria were able to grow and reproduce. This work was developed during World War II where in both Germany and France there was an increase in the application of anaerobic digestion for the treatment of manure.

Applications

Anaerobic digestion is particularly suited to wet organic material and is commonly used for effluent and sewage treatment. Anaerobic digestion is a simple process that can greatly reduce the amount of organic matter which might otherwise be destined to be landfilled or burnt in an incinerator.

Almost any organic material can be processed with anaerobic digestion. This includes biodegradable waste materials such as waste paper, grass clippings, leftover food, sewage and animal waste. The exception to this is woody wastes that are largely unaffected by digestion as most anaerobes are unable to degrade lignin. The exception being xylophalgeous anaerobes (lignin consumers), as used in the process for organic breakdown of cellulosic material by a cellulosic ethanol start-up company in the U.S. Anaerobic digesters can also be fed with specially grown energy crops such as silage for dedicated biogas production. In Germany and continental Europe these facilities are referred to as *biogas plants*. A *co-digestion* or *co-fermentation* plant is typically an agricultural anaerobic digester that accepts two or more input materials for simultaneous digestion.

In developing countries simple home and farm-based anaerobic digestion systems offer the potential for cheap, low-cost energy for cooking and lighting. Anaerobic digestion facilities have been recognized by the United Nations Development Programme as one of the most useful decentralized sources of energy supply. From 1975, China and India have both had large government-backed schemes for adaptation of small biogas plants for use in the household for cooking and lighting. Presently, projects for anaerobic digestion in the developing world can gain financial support through the United Nations Clean Development Mechanism if they are able to show they provide reduced carbon emissions.

Pressure from environmentally-related legislation on solid waste disposal methods in developed countries has increased the application of anaerobic digestion as a process for reducing waste volumes and generating useful by-products. Anaerobic digestion may either be used to process the source separated fraction of municipal waste, or alternatively combined with mechanical sorting systems, to process residual mixed municipal waste. These facilities are called mechanical biological treatment plants.

Utilising anaerobic digestion technologies can help to reduce the emission of greenhouse gasses in a number of key ways:

- Replacement of fossil fuels
- Reducing methane emission from landfills
- Displacing industrially-produced chemical fertilizers
- Reducing vehicle movements
- Reducing electrical grid transportation losses

Methane and power produced in anaerobic digestion facilities can be utilized to replace energy derived from fossil fuels, and hence reduce emissions of greenhouse gasses. This is due to the fact that the carbon in biodegradable material is part of a carbon cycle. The carbon released into the atmosphere from the combustion of biogas has been removed by plants in order for them to grow in the recent past. This can have occurred within the last decade, but more typically within the last growing season. If the plants are re-grown, taking the carbon out of the atmosphere once more, the system will be carbon neutral. This contrasts to carbon in fossil fuels that has been sequestered in the earth for many millions of years, the combustion of which increases the overall levels of carbon dioxide in the atmosphere.

If the putrescible waste processed in anaerobic digesters was disposed of in a landfill, it would break down naturally and often anaerobically. In this case the gas will eventually escape into the atmosphere. As methane is about twenty times more potent as a greenhouse gas as carbon dioxide this has significant negative environmental effects.

Digestate liquor can be used as a fertilizer supplying vital nutrients to soils. The solid, fibrous component of digestate can be used as a soil conditioner. The liquor can be used as a substitute for chemical fertilizers which require large amounts of energy to produce and transport. The use of manufactured fertilizers is therefore more carbon intensive than the use of anaerobic digestate fertiliser. This solid digestate can be used to boost the organic content of soils. There are some countries, such as Spain where there are many organically depleted soils, and here the markets for the digestate can be just as important as the biogas.

In countries that collect household waste, the utilization of local anaerobic digestion facilities can help to reduce the amount of waste that requires transportation to centralized landfill sites or incineration facilities. This reduced burden on transportation has and will reduce carbon emissions from the collection vehicles. If localized anaerobic digestion facilities are embedded within an electrical distribution network, they can help reduce the electrical losses that are associated with transporting electricity over a national grid.

In Oakland, California at the East Bay Municipal Utility District's (EBMUD) Main Wastewater Treatment Plant(MWWTP), food waste is currently co-digested with primary and secondary municipal wastewater solids and other high-strength wastes. Compared to municipal wastewater solids digestion, food waste digestion has many benefits. Anaerobic digestion of food waste pulp from the EBMUD food waste process provides a higher normalized energy benefit, compared to municipal wastewater solids:

- 730 to 1,300 kWh per dry ton of food waste applied.
- 560 to 940 kWh per dry ton of municipal wastewater solids applied.

POWER GENERATION

Biogas is usually used to run a CHP gas engine to produce electrical power some or all of which can be used to run the sewage works. The waste heat from the engine is then used to heat the dig ester. It turns out that the waste heat is generally enough to heat the dig ester to the required temperatures. The power potential from sewage works is trivial - in the UK there are about 80MW total of such generation, and this could be conceivably increased to 150 MW, which is insignificant compared to the average power demand in the UK of about 35,000 MW. The scope for biogas generation from non-sewage waste biological matter - food waste, abattoir waste etc is much higher, estimated to be capable of about 3,000MW.

The Process

There are a number of bacteria that are involved in the process of anaerobic digestion including acetic acid-forming bacteria (acetogens) and methane-forming archaea (methanogens). These organisms feed upon the initial feedstock, which undergoes a number of different processes converting it to intermediate molecules including sugars, hydrogen & acetic acid before finally being converted to biogas.

Different species of bacteria are able to survive at different temperature ranges. Ones living optimally at temperatures between 35-40°C are called mesophiles or mesophilic bacteria.

Some of the bacteria can survive at the hotter and more hostile conditions of 55-60°C, these are called thermophiles or thermophilic bacteria. Methanogens come from the primitive group of archaea. This family includes species that can grow in the hostile conditions of hydrothermal vents. These species are more resistant to heat and can therefore operate at thermophilic temperatures, a property that is unique to bacterial families.

As with aerobic systems the bacteria in anaerobic systems the growing and reproducing microorganisms within them require a source of elemental oxygen to survive.

Stages

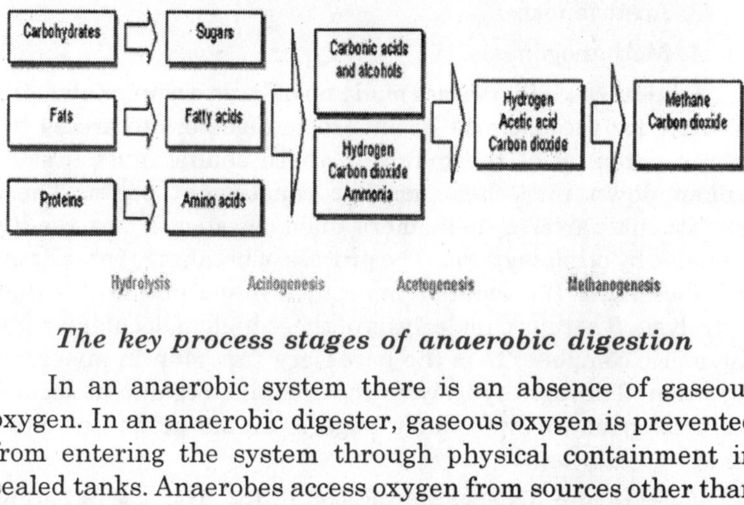

The key process stages of anaerobic digestion

In an anaerobic system there is an absence of gaseous oxygen. In an anaerobic digester, gaseous oxygen is prevented from entering the system through physical containment in sealed tanks. Anaerobes access oxygen from sources other than the surrounding air. The oxygen source for these microorganisms can be the organic material itself or alternatively may be supplied by inorganic oxides from within the input material. When the oxygen source in an anaerobic system is derived from the organic material itself, then the 'intermediate' end products are primarily alcohols, aldehydes, and organic acids plus carbon dioxide. In the presence of specialised methanogens, the intermediates are converted to the 'final' end products of methane, carbon dioxide with trace levels of hydrogen sulfide. In an anaerobic system the majority of the chemical energy contained within the starting material is released by methanogenic bacteria as methane.

Populations of anaerobic microorganisms typically take a significant period of time to establish themselves to be fully effective. It is therefore common practice to introduce anaerobic microorganisms from materials with existing populations. This process is called 'seeding' the digesters and typically takes place with the addition of sewage sludge or cattle slurry.

There are four key biological and chemical stages of anaerobic digestion:

1. Hydrolysis
2. Acidogenesis
3. Acetogenesis
4. Methanogenesis

In most cases biomass is made up of large organic polymers. In order for the bacteria in anaerobic digesters to access the energy potential of the material, these chains must first be broken down into their smaller constituent parts. These constituent parts or monomers such as sugars are readily available by other bacteria. The process of breaking these chains and dissolving the smaller molecules into solution is called hydrolysis. Therefore hydrolysis of these high molecular weight polymeric components is the necessary first step in anaerobic digestion. Through hydrolysis the complex organic molecules are broken down into simple sugars, amino acids, and fatty acids.

Acetate and hydrogen produced in the first stages can be used directly by methanogens. Other molecules such as volatile fatty acids (VFA's) with a chain length that is greater than acetate must first be catabolised into compounds that can be directly utilised by methanogens.

The biological process of acidogenesis is where there is further breakdown of the remaining components by acidogenic (fermentative) bacteria. Here VFAs are created along with ammonia, carbon dioxide and hydrogen sulfide as well as other by-products. The process of acidogenesis is similar to the way that milk sours.

The third stage anaerobic digestion is acetogenesis. Here simple molecules created through the acidogenesis phase are further digested by acetogens to produce largely acetic acid as well as carbon dioxide and hydrogen.

The terminal stage of anaerobic digestion is the biological process of methanogenesis. Here methanogens utilise the intermediate products of the preceding stages and convert them into methane, carbon dioxide and water. It is these components that makes up the majority of the biogas emitted from the system. Methanogenesis is sensitive to both high and low pHs and occurs between pH 6.5 and pH 8. The remaining, non-digestable material which the microbes cannot feed upon, along with any dead bacterial remains constitutes the digestate.

A simplified generic chemical equation for the overall processes outlined above is as follows:

$$C_6H_{12}O_6 \rightarrow 3CO_2 + 3CH_4$$

Feedstock

The most important initial issue when considering the application of anaerobic digestion systems is the feedstock to the process. Digesters typically can accept any biodegradable material, however if biogas production is the aim, the level of putrescibility is the key factor in its successful application. The more putrescible the material the higher the gas yields possible from the system.

Substrate composition is a major factor in determining the methane yield and methane production rates from the digestion of biomass. Techniques are available to determine the compositional characteristics of the feedstock, whilst parameters such as solids, elemental and organic analyses are important for digester design and operation.

Anaerobes can breakdown material to varying degrees of success from readily in the case of short chain hydrocarbons such as sugars, to over longer periods of time in the case of cellulose and hemicellulose. Anaerobic microorganisms are unable to break down long chain woody molecules such as lignin. Anaerobic digesters were originally designed for operation using sewage sludge and manures. Sewage and manure are not, however, the material with the most potential for anaerobic digestion as the biodegradable material has already had the energy content taken out by the animal that produced it.

A second consideration related to the feedstock will be moisture content. The wetter the material the more suitable it will be to handling with standard pumps instead of energy intensive concrete pumps and physical means of movement. Also the wetter the material, the more volume and area it takes up relative to the levels of gas that are produced. The moisture content of the target feedstock will also affect what type of system is applied to its treatment. In order to use a high solids anaerobic digester for dilute feedstocks, bulking agents such as compost should be applied to increase the solid content of the input material. Another key consideration is the carbon:nitrogen ratio of the input material. This ratio is the balance of food a microbe requires in order to grown. The optimal C:N ratio for the 'food' a microbe is 20–30:1.

The level of contamination of the feedstock material is a key consideration. If the feedstock to the digesters has significant levels of physical contaminants such as plastic, glass or metals then pre-processing will be required in order for the material to be used. If it is not removed then the digesters can be blocked and will not function efficiently. It is with this logic in mind that mechanical biological treatment plants are designed. The higher the level of pre-treatment a feedstock requires, the more processing machinery will be required and hence the project will have higher capital costs.

After sorting or screening to remove any physical contaminants, such as metals and plastics, from the feedstock the material is often shredded, minced and mechanically or hydraulically pulped to increase the surface area available to microbes in the digesters and hence increase the speed of digestion. The feedstock material is then fed into the airtight digester where anaerobic treatment takes place.

Configuration

Anaerobic digesters can be designed and engineered to operate using a number of different process configurations:

- Batch or continuous
- Temperature: Mesophilic or thermophilic
- Solids content: High solids or low solids
- Complexity: Single stage or multistage

Batch or Continuous

A batch system is the simplest form of digestion. Biomass is added to the reactor at the start of the process in a batch and is sealed for the duration of the process. Batch reactors suffer from odour issues that can be a severe problem when they are emptied. Typically biogas production will be formed with a normal distribution pattern over time. The operator can use this fact to determine when they believe the process of digestion of the organic matter has completed. As the batch digestion is simple and requires less equipment and lower levels of design work it is typically a cheaper form of digestion.

In continuous digestion processes organic matter is constantly or added in stages to the reactor. Here the end products are constantly or periodically removed, resulting in constant production of biogas. Examples of this form of anaerobic digestion include, continuous stirred-tank reactors (CSTRs), Upflow anaerobic sludge blanket (UASB), Expanded granular sludge bed (EGSB) and Internal circulation reactors (IC).

Temperature

There are two conventional operational temperature levels for anaerobic digesters, which are determined by the species of methanogens in the digesters:

- *Mesophilic* which takes place optimally around 37°-41°C or at ambient temperatures between 20°-45°C where mesophiles are the primary microorganism present
- *Thermophilic* which takes place optimally around 50°-52° at elevated temperatures up to 70°C where thermophiles are the primary microorganisms present

There are a greater number of species of mesophiles than thermophiles. These bacteria are also more tolerant to changes in environmental conditions than thermophiles. Mesophilic systems are therefore considered to be more stable than thermophilic digestion systems.

As mentioned above, thermophilic digestion systems are considered to be less stable, however the increased temperatures facilitate faster reaction rates and hence faster gas yields. Operation at higher temperatures facilitates greater sterilisation of the end digestate. In countries where legislation, such as the Animal By-Products Regulations in the European Union, requires end products to meet certain levels of reduction in the amount of bacteria in the output material, this may be a benefit.

A drawback of operating at thermophilic temperatures is that more heat energy input is required to achieve the correct operational temperatures. This increase in energy may not be outweighed by the increase in the outputs of biogas from the systems. It is therefore important to consider an energy balance for these systems.

Solids

Typically there are two different operational parameters associated with the solids content of the feedstock to the digesters:

- High-solids
- Low-solids

Digesters can either be designed to operate in a high solids content, with a total suspended solids (TSS) concentration greater than ~20%, or a low solids concentration less than ~15%.

High-solids digesters process a thick slurry that requires more energy input to move and process the feedstock. The thickness of the material may also lead to associated problems with abrasion. High-solids digesters will typically have a lower land requirement due to the lower volumes associated with the moisture.

Low-solids digesters can transport material through the system using standard pumps that require significantly lower energy input. Low-solids digesters require a larger amount of land than high-solids due to the increase volumes associated with the increased liquid-to-feedstock ratio of the digesters. There are benefits associated with operation in a liquid environment as it enables more thorough circulation of materials and contact between the bacteria and their food. This enables the bacteria to more readily access the substances they are feeding off and increases the speed of gas yields.

Number of Stages

Digestion systems can be configured with different levels of complexity:

- One-stage or single-stage
- Two-stage or multistage

A single-stage digestion system is one in which all of the biological reactions occur within a single sealed reactor or holding tank. Utilising a single stage reduces construction costs, however facilitates less control of the reactions occurring within the system. Acidogenic bacteria, through the production of acids, reduce the pH of the tank. Methanogenic bacteria, as outlined earlier, operate in a strictly defined pH range. Therefore the biological reactions of the different species in a single stage reactor can be in direct competition with each other. Another one-stage reaction system is an anaerobic lagoon. These lagoons are pond-like earthen basins used for the treatment and long-term storage of manures. Here the anaerobic reactions are contained within the natural anaerobic sludge contained in the pool.

In a two-stage or multi-stage digestion system different digestion vessels are optimised to bring maximum control over the bacterial communities living within the digesters. Acidogenic bacteria produce organic acids and more quickly grow and reproduce than methanogenic bacteria. Methanogenic bacteria require stable pH and temperature in order to optimise their performance.

Typically hydrolysis, acetogenesis and acidogenesis occur within the first reaction vessel. The organic material is then heated to the required operational temperature (either mesophilic or thermophilic) prior to being pumped into a methanogenic reactor. The initial hydrolysis or acidogenesis tanks prior to the methanogenic reactor can provide a buffer to the rate at which feedstock is added. Some European countries require a degree of elevated heat treatment in order to kill harmful bacteria in the input waste. In this instance there may be a pasteurisation or sterilisation stage prior to digestion or between the two digestion tanks. It should be noted that it is not possible to completely isolate the different reaction phases and often there is some biogas that is produced in the hydrolysis or acidogenesis tanks.

Residence

The residence time in a digester varies with the amount and type of feed material, the configuration of the digestion system and whether it be one-stage or two-stage.

In the case of single-stage thermophilic digestion residence times may be in the region of 14 days, which comparatively to mesophilic digestion is relatively fast. The plug-flow nature of some of these systems will mean that the full degradation of the material may not have been realised in this timescale. In this event digestate exiting the system will be darker in colour and will typically have more odour.

In two-stage mesophilic digestion, residence time may vary between 15 and 40 days. In the case of mesophilic UASB digestion hydraulic residence times can be (1hour-1day) and solid retention times can be up to 90 days. In this manner the UASB system is able to separate solid an hydraulic retention times with the utilisation of a sludge blanket.

Continuous digesters have mechanical or hydraulic devices, depending on the level of solids in the material, to mix the contents enabling the bacteria and the food to be in contact. They also allow excess material to be continuously extracted to maintain a reasonably constant volume within the digestion tanks.

Products

There are three principal products of anaerobic digestion: biogas, digestate and water.

Biogas

Typical composition of biogas

Matter	%
Methane, CH_4	50-75
Carbon dioxide, CO_2	25-50
Nitrogen, N_2	0-10
Hydrogen, H_2	0-1
Hydrogen sulfide, H_2S	0-3
Oxygen, O_2	0-2

Biogas is the ultimate waste product of the bacteria feeding off the input biodegradable feedstock, and is mostly methane and carbon dioxide, with a small amount hydrogen and trace hydrogen sulfide. Most of the biogas is produced during the middle of the digestion, after the bacterial population has grown, and tapers off as the putrescible material is exhausted. The gas is normally stored on top of the digester in an inflatable gas bubble or extracted and stored next to the facility in a gas holder. The methane in biogas can be burned to produce both heat and electricity, usually with a reciprocating engine or microturbine often in a cogeneration arrangement where the electricity and waste heat generated are used to warm the digesters or to heat buildings. Excess electricity can be sold to suppliers or put into the local grid. Electricity produced by anaerobic digesters is considered to be renewable energy and may attract subsidies. Biogas does not contribute to increasing atmospheric carbon dioxide concentrations because the gas is not released directly into the atmosphere and the carbon dioxide comes from an organic source with a short carbon cycle.

Biogas may require treatment or 'scrubbing' to refine it for use as a fuel. Hydrogen sulfide is a toxic product formed from sulfates in the feedstock and is released as a trace component of the biogas. National environmental enforcement agencies such as the U.S. Environmental Protection Agencyý or the English and Welsh Environment Agency put strict limits on the levels of gasses containing hydrogen sulfide, and if the levels of hydrogen sulfide in the gas are high, gas scrubbing and cleaning equipment (such as amine gas treating) will be needed to process the biogas to within regionally accepted levels. An alternative method to this is by the addition of ferric chloride $FeCl_3$ to the digestion tanks in order to inhibit hydrogen sulfide production. Volatile siloxanes can also contaminate the biogas; such compounds are frequently found in household waste and wastewater. In digestion facilities accepting these materials as a component of the feedstock, low molecular weight siloxanes volatilise into biogas. When this gas is combusted in a gas engine, turbine or boiler, siloxanes are converted into silicon dioxide ($Si0_2$) which deposits internally in the machine, increasing wear and tear. In countries such as Switzerland, Germany and Sweden the methane in the biogas may be concentrated in order for it to be used as a vehicle transportation fuel or alternatively input directly into the gas mains. In countries where the driver for the utilisation of anaerobic digestion are renewable electricity subsidies, this route of treatment is less likely as energy is required in this processing stage and reduces the over all levels available to sell.

Digestate

Digestate is the solid remnants of the original input material to the digesters that the microbes cannot use. It also consists of the mineralised remains of the dead bacteria from within the digesters. Digestate can come in three forms; fibrous, liquor or a sludge-based combination of the two fractions. In two-stage systems the different forms of digestate come from different digestion tanks. In single stage digestion systems the two fractions will be combined and if desired separated by further processing.

The second by-product (acidogenic digestate) is a stable organic material comprised largely of lignin and cellulose, but also of a variety of mineral components in a matrix of dead bacterial cells; some plastic may be present. The material resembles domestic compost and can be used as compost or to make low grade building products such as fibreboard. The third by-product is a liquid (methanogenic digestate) that is rich in nutrients and can be used as a fertiliser dependent on the quality of the material being digested. Levels of potentially toxic elements (PTEs) should be chemically assessed. This will be dependent upon the quality of the original feedstock. In the case of most clean and source-separated biodegradable waste streams the levels of PTEs will be low. In the case of wastes originating from industry the levels of PTEs may be higher and will need to be taken into consideration when determining a suitable end use for the material. Digestate typically contains elements such as lignin that cannot be broken down by the anaerobic microorganisms. Also the digestate may contain ammonia that is phytotoxic and will hamper the growth of plants if it is used as a soil improving material. For these two reasons a maturation or composting stage may be employed after digestion. Lignin and other materials are available for degradation by aerobic microorganisms such as fungi helping reduce the overall volume of the material for transport. During this maturation the ammonia will be broken down into nitrates, improving the fertility of the material and making it more suitable as a soil improver. Large composting stages are typically used by dry anaerobic digestion technologies.

Wastewater

The final output from anaerobic digestion systems is water. This water originates both from the moisture content of the original waste that was treated but also includes water produced during the microbial reactions in the digestion systems. This water may be released from the dewatering of the digestate or may be implicitly separate from the digestate.

The wastewater exiting the anaerobic digestion facility will typically have elevated levels of biochemical oxygen demand (BOD) and chemical oxygen demand (COD), these are measures of the reactivity of the effluent and show an ability to pollute. Some of this material is termed 'hard COD' meaning it cannot be accessed by the anaerobic bacteria for conversion into biogas. If this effluent was put directly into watercourses it would negatively affect them by causing eutrophication. As such further treatment of the wastewater is often required. This treatment will typically be an oxidation stage where air is passed through the water in a sequencing batch reactors or reverse osmosis unit.

6

Renewable Energy Commercialization

Renewable energy commercialization involves the diffusion of three generations of technologies dating back more than 100 years. First-generation technologies, which are already mature and economically competitive, include biomass, hydroelectricity, geothermal power and heat. Second-generation technologies are market-ready and are being deployed at the present time; they include solar heating, photovoltaics, wind power, solar thermal power stations, and modern forms of bioenergy. Third-generation technologies require continued R&D efforts in order to make large contributions on a global scale and include advanced biomass gasification, biorefinery technologies, hot-dry-rock geothermal power, and ocean energy.

While there are many non-technical barriers to the widespread use of renewables, some 65 countries now have targets for their own renewable energy futures, and have enacted wide-ranging public policies to promote renewables. Climate change concerns are driving increasing growth in the renewable energy industries and investment capital flowing into renewable energy reached a record US$77 billion in 2007. Leading renewable energy companies include: Enercon, Gamesa, GE Energy, Q-Cells, Sharp Solar, SunOpta, and Vestas.

Global revenues for solar photovoltaics, wind power, and biofuels expanded from $76 billion in 2007 to $115 billion in 2008. New global investments in clean energy technologies — including venture capital, project finance, public markets, and research and development — expanded by 4.7 percent from

$148 billion in 2007 to $155 billion in 2008. Continued growth for the renewable energy sector is expected in the mid- to long-term, but 2009 will be a year of refocus, consolidation, or retrenchment for many companies. At the same time, new government spending, regulation, and policies should help the industry weather the current economic crisis better than many other sectors. Most notably, U.S. President Barack Obama's American Recovery and Reinvestment Act of 2009 includes more than $70 billion in direct spending and tax credits for clean energy and associated transportation programs. Clean Edge suggests that the commercialization of clean energy will help countries around the world pull out of the current economic malaise.

OVERVIEW

The International Energy Agency (IEA) has defined three generations of renewable energy technologies, reaching back over 100 years:

- *First-generation technologies* emerged from the industrial revolution at the end of the 19th century and include hydropower, biomass combustion, geothermal power and heat. These technologies are quite widely used.
- *Second-generation technologies* include solar heating and cooling, wind power, modern forms of bioenergy, and solar photovoltaics. These are now entering markets as a result of research, development and demonstration (RD&D) investments since the 1980s. Initial investment was prompted by energy security concerns linked to the oil crises of the 1970s but the enduring appeal of these technologies is due, at least in part, to environmental benefits. Many of the technologies reflect significant advancements in materials.
- *Third-generation technologies* are still under development and include advanced biomass gasification, biorefinery technologies, concentrating solar thermal power, hot-dry-rock geothermal power, and ocean energy. Advances in nanotechnology may also play a major role.

First-generation technologies are well established, second-generation technologies are entering markets, and third-generation technologies heavily depend on long-term RD&D commitments, where the public sector has a role to play.

Renewable energy technologies are essential contributors to the energy supply portfolio, as they contribute to world energy security, reduce dependency on fossil fuels, and provide opportunities for mitigating greenhouse gases. The IEA estimates that nearly 50% of global electricity supplies will need to come from renewable energy sources in order to halve CO_2 emissions by 2050 and minimise significant, irreversible climate change impacts.

FIRST-GENERATION TECHNOLOGIES

First-generation technologies are widely used in locations with abundant resources. Their future use depends on the exploration of the remaining resource potential, particularly in developing countries, and on overcoming challenges related to the environment and social acceptance.

Biomass

Biomass for heat and power is a fully mature technology which offers a ready disposal mechanism for municipal, agricultural, and industrial organic wastes. However, the industry has remained relatively stagnant over the decade to 2007, even though demand for biomass (mostly wood) continues to grow in many developing countries. One of the problems of biomass is that material directly combusted in cook stoves produces pollutants, leading to severe health and environmental consequences, although improved cook stove programmes are alleviating some of these effects. First-generation biomass technologies can be economically competitive, but may still require deployment support to overcome public acceptance and small-scale issues.

Hydroelectricity

Hydroelectric plants have the advantage of being long-lived and many existing plants have operated for more than 100 years. Hydropower is also an extremely flexible technology from the perspective of power grid operation. Large hydropower provides one of the lowest cost options in today's energy market, even compared to fossil fuels and there are no harmful emissions associated with plant operation.

However, there are several significant social and environmental disadvantages of large-scale hydroelectric power systems: dislocation of people living where the reservoirs are planned, release of significant amounts of carbon dioxide and methane during construction and flooding of the reservoir, and disruption of aquatic ecosystems and birdlife. Hydroelectric power is now more difficult to site in developed nations because most major sites within these nations are either already being exploited or may be unavailable for these environmental reasons. The areas of greatest hydroelectric growth are the growing economies of Asia. China is the development leader; however, other Asian nations are also expanding hydropower.

There is a strong consensus now that countries should adopt an integrated approach towards managing water resources, which would involve planning hydropower development in co-operation with other water-using sectors.

GEOTHERMAL POWER AND HEAT

Geothermal power plants can operate 24 hours per day, providing baseload capacity, and the world potential capacity for geothermal power generation is estimated at 85 GW over the next 30 years. However, geothermal power is accessible only in limited areas of the world. The costs of geothermal energy have dropped substantially from the systems built in the 1970s.

Geothermal heat generation can be competitive in many countries producing geothermal power, or in other regions where the resource is of a lower temperature.

SECOND-GENERATION TECHNOLOGIES

Markets for second-generation technologies have been strong and growing over the past decade, and these technologies have gone from being a passion for the dedicated few to a major economic sector in countries such as Germany, Spain, the United

States, and Japan. Many large industrial companies and financial institutions are involved and the challenge is to broaden the market base for continued growth worldwide.

Solar Heating

Solar heating systems are a well known second-generation technology and generally consist of solar thermal collectors, a fluid system to move the heat from the collector to its point of usage, and a reservoir or tank for heat storage. The systems may be used to heat domestic hot water, swimming pools, or homes and businesses. The heat can also be used for industrial process applications or as an energy input for other uses such as cooling equipment. In many warmer climates, a solar heating system can provide a very high percentage (50 to 75%) of domestic hot water energy. An early solar heating boom took place during the 1940s in the United States, during which period institutional support for solar research and energy conservation measures imposed during World War II fueled significant advances in solar technology, which went as far as the development of a prototype prefabricated solar-heated home. A few proponents of this technology saw it as a clean alternative to polluting fuels, but the great majority of advocates, researchers, and investors saw it as a solution to high energy costs during the war; when those conditions changed and the 1950s ushered in a period of record low energy prices, interest rapidly waned, and the commercial development of solar heating systems was postponed to a later decade.

Photovoltaics

Photovoltaic (PV) cells, also called solar cells, convert light into electricity. In the 1980s and early 1990s, most photovoltaic modules were used to provide Remote Area Power Supply, but from around 1995, industry efforts have focused increasingly on developing building integrated photovoltaics and photovoltaic power stations for grid connected applications. Currently the largest photovoltaic power plant in North America is the Nellis Solar Power Plant (15 MW). There is a proposal to build a Solar power station in Victoria, Australia, which would be the world's largest PV power station, at 154 MW. Other large photovoltaic power stations, which are under construction, include the Girassol solar power plant (62 MW), and the Waldpolenz Solar Park (40 MW). Annual production of photovoltaics reached 3,800 megawatts worldwide in 2007, an increase of 50 percent over 2006. At the end of 2007, according to preliminary data, cumulative global production was 12,400 megawatts. Photovoltaic production has been doubling every two years, increasing by an average of 48 percent each year since 2002, making it the world's fastest-growing energy technology. The top five photovoltaic producing countries are Japan, China, Germany, Taiwan, and the USA.

WIND POWER

Some of the second-generation renewables, such as wind power, have high potential and have already realised relatively low production costs. At the end of 2008, worldwide wind farm capacity was 120,791 megawatts (MW), representing an increase of 28.8 percent during the year, and wind power produced some 1.3% of global electricity consumption. Wind power accounts for approximately 19% of electricity use in Denmark, 9% in Spain and Portugal, and 6% in Germany and the Republic of Ireland. However, it may be difficult to site wind turbines in some areas for aesthetic or environmental reasons.

The United States is an important growth area and installed U.S. wind power capacity reached 25,170 MW at the end of 2008. Some of the largest wind farms operating in the U.S. are: Horse Hollow Wind Energy Center, TX (736 MW); Maple Ridge Wind Farm, NY (322 MW); Stateline Wind Project, OR & WA (300 MW); King Mountain Wind Farm, TX (281 MW); and Sweetwater Wind Farm, TX (264 MW).

SOLAR THERMAL POWER STATIONS

Solar thermal power stations include the 354 MW Solar Energy Generating Systems power plant in the USA, Nevada Solar One (USA, 64 MW), Andasol 1 (Spain, 50 MW) and the PS10 solar power tower (Spain, 11 MW). Many other plants are under construction or planned, mainly in Spain and the USA. In developing countries, three World Bank projects for integrated solar thermal/combined-cycle gas-turbine power plants in Egypt, Mexico, and Morocco have been approved.

Modern Forms of Bioenergy

Brazil has one of the largest renewable energy programs in the world, involving production of ethanol fuel from sugar cane, and ethanol now provides 18 percent of the country's automotive fuel. As a result of this and the exploitation of domestic deep water oil sources, Brazil, which for years had to import a large share of the petroleum needed for domestic consumption, recently reached complete self-sufficiency in liquid fuels.

Production and use of ethanol has been stimulated through: (1) low-interest loans for the construction of ethanol distilleries; (2) guaranteed purchase of ethanol by the state-owned oil company at a reasonable price; (3) retail pricing of neat ethanol so it is competitive if not slightly favorable to the gasoline-ethanol blend; and (4) tax incentives provided during the 1980s to stimulate the purchase of neat ethanol vehicles. Guaranteed purchase and price regulation were ended some years ago, with relatively positive results. In addition to these other policies, ethanol producers in the state of São Paulo established a research and technology transfer center that has been effective in improving sugar cane and ethanol yields.

Most cars on the road today in the U.S. can run on blends of up to 10% ethanol, and motor vehicle manufacturers already produce vehicles designed to run on much higher ethanol blends. Ford, DaimlerChrysler, and GM are among the automobile companies that sell "flexible-fuel" cars, trucks, and minivans that can use gasoline and ethanol blends ranging from pure gasoline up to 85% ethanol (E85). By mid-2006, there were approximately six million E85-compatible vehicles on U.S. roads. The challenge is to expand the market for biofuels beyond the farm states where they have been most popular to date. Flex-fuel vehicles are assisting in this transition because they allow drivers to choose different fuels based on price and availability. The Energy Policy Act of 2005, which calls for 7.5 billion gallons of biofuels to be used annually by 2012, will also help to expand the market. It should also be noted that the growing ethanol and biodiesel industries are providing jobs in plant construction, operations, and maintenance, mostly in rural communities. According to the Renewable Fuels Association, the ethanol industry created almost 154,000 U.S. jobs in 2005 alone, boosting household income by $5.7 billion. It also contributed about $3.5 billion in tax revenues at the local, state, and federal levels.

Third-generation Technologies

Third-generation renewable energy technologies are still under development and include advanced biomass gasification, biorefinery technologies, solar thermal power stations, hot-dry-rock geothermal power, and ocean energy. Third-generation technologies are not yet widely demonstrated or have limited commercialization. Many are on the horizon and may have potential comparable to other renewable energy technologies, but still depend on attracting sufficient attention and RD&D funding.

New Bioenergy Technologies

Commercial Cellulosic Ethanol Plants in the U.S. (Operational or under construction)

Company	Location	Feedstock
Abengoa Bioenergy	Hugoton, KS	Wheat straw
BlueFire Ethanol	Irvine, CA	Multiple sources
Colusa Biomass Energy Corporation	Sacramento, CA	Waste rice straw
Fulcrum BioEnergy	Reno, NV	Municipal solid waste
Gulf Coast Energy	Mossy Head, FL	Wood waste
KL Energy Corp.	Upton, WY	Wood
Mascoma	Lansing, MI	Wood
POET LLC	Emmetsburg, IA	Corn cobs
Range Fuels	Treutlen County, GA	Wood waste
SunOpta	Little Falls, MN	Wood chips
US Envirofuels	Highlands County, FL	Sweet sorghum
Xethanol	Auburndale, FL	Citrus peels

According to the International Energy Agency, cellulosic ethanol biorefineries could allow biofuels to play a much bigger role in the future than organizations such as the IEA previously thought. Cellulosic ethanol can be made from plant matter composed primarily of inedible cellulose fibers that form the stems and branches of most plants. Crop residues (such as corn stalks, wheat straw and rice straw), wood waste, and municipal solid waste are potential sources of cellulosic biomass. Dedicated energy crops, such as switchgrass, are also promising cellulose sources that can be sustainably produced in many regions of the United States.

OCEAN ENERGY

In terms of ocean energy, another third-generation technology, Portugal has the world's first commercial wave farm, the *Aguçadora Wave Park*, opened in 2008. The first stage of the farm uses three Pelamis P-750 machines generating a total of 2.25 MW. The cost of the farm is put at 8.5 million euro. A second phase of the project is now planned to increase the installed capacity from 2.25 MW to 21 MW using a further 25 Pelamis machines. Funding for a wave farm in Scotland was announced in February 2007 by the Scottish Executive, at a cost of over 4 million pounds, as part of a £13 million funding packages for ocean power in Scotland. The farm will be the world's largest with a capacity of 3 MW generated by four Pelamis machines. In 2007, the world's first commercial tidal power station was installed in the narrows of Strangford Lough in Ireland. The 1.2 megawatt underwater tidal electricity generator, part of Northern Ireland's Environment & Renewable Energy Fund scheme, takes advantage of the fast tidal flow (up to 4 metres per second) in the lough. Although the generator is powerful enough to power a thousand homes, the turbine has minimal environmental impact, as it is almost entirely submerged, and the rotors pose no danger to wildlife as they turn quite slowly.

ENHANCED GEOTHERMAL SYSTEMS

Enhanced geothermal systems, also known as hot dry rock geothermal, utilise new techniques to exploit resources that would have been uneconomical in the past. These systems are still in the research phase, and require additional R&D for new and improved approaches, as well as to develop smaller modular units that will allow economies of scale at the manufacturing level. Further government-funded research and close collaboration with industry will help to make exploitation of geothermal resources more economically attractive for investors.

NANOTECHNOLOGY THIN-FILM SOLAR PANELS

Solar power panels that use nanotechnology, which can create circuits out of individual silicon molecules, may cost half as much as traditional photovoltaic cells, according to executives and investors involved in developing the products. Nanosolar has secured more than $100 million from investors to build a factory for nanotechnology thin-film solar panels. The company expects the factory to open in 2010 and produce enough solar cells each year to generate 430 megawatts of power.

RENEWABLE ENERGY INDUSTRY

By mid-2007, some 140 publicly-traded renewable energy companies worldwide (or renewable energy divisions of major companies) each had a market capitalization greater than $40 million. The estimated total market capitalization of these companies and divisions was more than $100 billion in mid-2007.

In 2000, venture capital (VC) investment in renewable energy was about 1% of total VC investment. In 2007 that figure was closer to 10%, with solar power alone making up about 3% of the entire venture capital asset class of ~$33B. More than 60 start-ups have been funded by VCs in the last three years.

WIND POWER COMPANIES

Currently three quarters of global wind turbine sales come from only four turbine manufacturing companies: Vestas, Gamesa, Enercon, and GE Energy. Vestas is the largest wind turbine manufacturer in the world with a 28% market share. The company operates plants in Denmark, Germany, India,

Italy, Britain, Spain, Sweden, Norway, Australia and China, and employs more than 20,000 people globally. After a sales slump in 2005, Vestas recovered and was voted *Top Green Company of 2006*. Vestas announced a major expansion of its North American headquarters in Portland, Oregon in December, 2008. Gamesa, founded in 1976 with headquarters in Vitoria, Spain, is currently the world's second largest wind turbine manufacturer, after Vestas, and it is also a major builder of wind farms. Gamesa's main markets are within Europe, the US and China. In 2006, Europe accounted for 65 percent of Gamesa's sales, of which 40 percent were in Spain.

In 2004, German company Enercon installed a total of 1288 MW of wind power and had around 16% of the global market share. Enercon constructed production facilities in Brazil in 2006, and has extended its presence there, as well as in the more traditional markets of Germany, India, Austria, UK, Canada and the Netherlands.

GE Energy has installed over 5,500 wind turbines and 3,600 hydro turbines, and its installed capacity of renewable energy worldwide exceeds 160,000 MW. GE Energy bought out Enron Wind in 2002 and also has nuclear energy operations in its portfolio.

PHOTOVOLTAIC COMPANIES

Q-Cells became the world's largest solar cell maker in 2007, producing nearly 400 MW of product. Longtime market leader Sharp Corporation found itself in second place with production of 370 MW in 2007, which the company blamed on a constrained supply of silicon. China's Suntech was close behind the leaders with more than 300 MW of output. Kyocera and its 200 MW output was a distant fourth in 2007.

Four new companies entered the top ranks in 2007. CdTe-cell maker First Solar was at fifth place, the only US-based and only thin-film supplier in the Top 10 companies. Asian players Motech Solar (Taiwan), Yingli Green Energy (China), and JA Solar Holdings (China/Australia) rounded out the Top 10 ranking, pushing aside some established players like Mitsubishi Electric, Schott, and BP Solar.

OTHER COMPANIES

SunOpta is located in Canada and was founded in 1973. Its operations are divided between SunOpta Food (organics), Opta Minerals, and SunOpta BioProcess (bioethanol). SunOpta's fastest growing business segment is the BioProcess Group, which is a leading developer of technology in the cellulosic ethanol market. SunOpta's BioProcess Group specializes in the design, construction and optimization of biomass conversion equipment and facilities. They have over 30 years experience delivering biomass solutions worldwide and use innovative technologies to produce cellulosic ethanol and cellulosic butanol. Raw materials include wheat straw, corn stover, grasses, oat hulls and wood chips.

NON-TECHNICAL BARRIERS TO ACCEPTANCE

There have been several recent reports which have identified a range of "non-technical barriers" to renewable energy use. These barriers are impediments which put renewable energy at a marketing, institutional, or policy disadvantage relative to other forms of energy. Key barriers include:

- Lack of government policy support, which includes the lack of policies and regulations supporting deployment of renewable energy technologies and the presence of policies and regulations hindering renewable energy development and supporting conventional energy development. Examples include subsidies for fossil-fuels, insufficient consumer-based renewable energy incentives, government underwriting for nuclear plant accidents, and complex zoning and permitting processes for renewable energy.
- Lack of information dissemination and consumer awareness.
- Higher capital cost of renewable energy technologies compared with conventional energy technologies.
- Difficulty overcoming established energy systems, which includes difficulty introducing innovative energy systems, particularly for distributed generation such as photovoltaics, because of technological lock-in, electricity

markets designed for centralized power plants, and market control by established operators. As the Stern Review on the Economics of Climate Change points out:

National grids are usually tailored towards the operation of centralised power plants and thus favour their performance. Technologies that do not easily fit into these networks may struggle to enter the market, even if the technology itself is commercially viable. This applies to distributed generation as most grids are not suited to receive electricity from many small sources. Large-scale renewables may also encounter problems if they are sited in areas far from existing grids.

- Inadequate financing options for renewable energy projects, including insufficient access to affordable financing for project developers, entrepreneurs and consumers.
- Imperfect capital markets, which includes failure to internalize all costs of conventional energy (e.g., effects of air pollution, risk of supply disruption) and failure to internalize all benefits of renewable energy (e.g., cleaner air, energy security).
- Inadequate workforce skills and training, which includes lack of adequate scientific, technical, and manufacturing skills required for renewable energy production; lack of reliable installation, maintenance, and inspection services; and failure of the educational system to provide adequate training in new technologies.
- Lack of adequate codes, standards, utility interconnection, and net-metering guidelines.
- Poor public perception of renewable energy system aesthetics.
- Lack of stakeholder/community participation and co-operation in energy choices and renewable energy projects.

With such a wide range of non-technical barriers, there is no "silver bullet" solution to drive the transition to renewable energy. So ideally there is a need for several different types of policy instruments to complement each other and overcome different types of barriers. A policy framework must be created that will level the playing field and redress the imbalance of traditional approaches associated with fossil fuels. The policy landscape must keep pace with broad trends within the energy sector, as well as reflecting specific social, economic and environmental priorities.

EDUCATION AND RENEWABLE ENERGY

To deal with these non-technical barriers to acceptance of renewable energy and also to address growing need for knowhow on non fossil fuel energy and issues like climate changes, several universities have set up special courses or research programs on renewable energy. In some cases universities from different countries have joined forces to form network to promote renewable energy. RES - The School for Renewable Energy Science in Iceland is an example of such international cooperation.

Public Policy Landscape

Public policy has a role to play in renewable energy commercialization because the free market system has some fundamental limitations. As the Stern Review points out:

In a liberalised energy market, investors, operators and consumers should face the full cost of their decisions. But this is not the case in many economies or energy sectors. Many policies distort the market in favour of existing fossil fuel technologies. Lester Brown goes further and suggests that the market "does not incorporate the indirect costs of providing goods or services into prices, it does not value nature's services adequately, and it does not respect the sustainable-yield thresholds of natural systems". It also favors the near term over the long term, thereby showing limited concern for future generations. Tax and subsidy shifting can help overcome these problems.

Shifting Taxes

Tax shifting involves lowering income taxes while raising levies on environmentally destructive activities, in order to create a more responsive market. It has been widely discussed and endorsed by economists. For example, a tax on coal that included the increased health care costs associated with breathing polluted air, the costs of acid rain damage, and the costs of climate disruption would encourage investment in renewable technologies. Several Western European countries are already shifting taxes in a process known there as environmental tax reform, to achieve environmental goals.

A four-year plan adopted in Germany in 1999 gradually shifted taxes from labor to energy and, by 2001, this plan had lowered fuel use by 5 percent. It had also increased growth in the renewable energy sector, creating some 45,400 jobs by 2003 in the wind industry alone, a number that is projected to rise to 103,000 by 2010. In 2001, Sweden launched a new 10-year environmental tax shift designed to convert 30 billion kroner ($3.9 billion) of taxes on income to taxes on environmentally destructive activities. Other European countries with significant tax reform efforts are France, Italy, Norway, Spain, and the United Kingdom. Asia's two leading economies, Japan and China, are considering the adoption of carbon taxes.

Shifting Subsidies

Subsidies are not inherently bad as many technologies and industries emerged through government subsidy schemes. The Stern Review explains that of 20 key innovations from the past 30 years, only one of the 14 they could source was funded entirely by the private sector and nine were totally funded by the public sector. In terms of specific examples, the Internet was the result of publicly funded links among computers in government laboratories and research institutes. And the combination of the federal tax deduction and a robust state tax deduction in California helped to create the modern wind power industry.

But just as there is a need for tax shifting, there is also a need for subsidy shifting. Lester Brown has argued that "a world facing the prospect of economically disruptive climate change can no longer justify subsidies to expand the burning of coal and oil. Shifting these subsidies to the development of climate-benign energy sources such as wind, solar, biomass, and geothermal power is the key to stabilizing the earth's climate."

Some countries are eliminating or reducing climate disrupting subsidies and Belgium, France, and Japan have phased out all subsidies for coal. Germany reduced its coal subsidy from $5.4 billion in 1989 to $2.8 billion in 2002, and in the process lowered its coal use by 46 percent. Germany plans to phase out this support entirely by 2010. China cut its coal subsidy from $750 million in 1993 to $240 million in 1995 and more recently has imposed a tax on high-sulfur coals.

While some leading industrial countries have been reducing subsidies to fossil fuels, most notably coal, the United States has been increasing its support for the fossil fuel and nuclear industries.

RENEWABLE ENERGY TARGETS

Setting national renewable energy targets can be an important part of a renewable energy policy and these targets are usually defined as a percentage of the primary energy and/or electricity generation mix. For example, the European Union has prescribed an indicative renewable energy target of 12 per cent of the total EU energy mix and 22 per cent of electricity consumption by 2010. National targets for individual EU Member States have also been set to meet the overall target. Other developed countries with defined national or regional targets include Australia, Canada, Japan, New Zealand, Norway, Switzerland, and some US States.

National targets are also an important component of renewable energy strategies in some developing countries. Developing countries with renewable energy targets include China, India, Korea, Indonesia, Malaysia, the Philippines, Singapore, Thailand, Brazil, Israel, Egypt, Mali, and South Africa. The targets set by many developing countries are quite modest when compared with those in some industrialized countries.

Renewable energy targets in most countries are indicative and nonbinding but they have assisted government actions and regulatory frameworks. The United Nations Environment

Program has suggested that making renewable energy targets legally binding could be an important policy tool to achieve higher renewable energy market penetration.

Employment

Current employment in the renewable energy sector and supplier industries is estimated at 2.3 million worldwide. The wind power industry employs some 300,000 people, the photovoltaics sector an estimated 170,000, and the solar thermal industry more than 600,000. Over 1 million jobs are found in the biofuels industry, associated with growing and processing a variety of feedstocks into ethanol and biodiesel.

Recent Developments

A number of events in 2006 pushed renewable energy up the political agenda, including the US mid-term elections in November, which confirmed clean energy as a mainstream issue. Also in 2006, the Stern Review made a strong economic case for investing in low carbon technologies now, and argued that economic growth need not be incompatible with cutting energy consumption. According to a trend analysis from the United Nations Environment Programme, climate change concerns coupled with recent high oil prices and increasing government support are driving increasing rates of investment in the renewable energy and energy efficiency industries.

Investment capital flowing into renewable energy reached a record US$77 billion in 2007, with the upward trend continuing in 2008. The OECD still dominates, but there is now increasing activity from companies in China, India and Brazil. Chinese companies were the second largest recipient of venture capital in 2006 after the United States. In the same year, India was the largest net buyer of companies abroad, mainly in the more established European markets.

Global revenues for solar photovoltaics, wind power, and biofuels expanded from $75.8 billion in 2007 to $115.9 billion in 2008. For the first time, one sector alone, wind, had revenues exceeding $50 billion. New global investments in clean energy technologies — including venture capital, project finance, public markets, and research and development — expanded by 4.7

percent from $148.4 billion in 2007 to $155.4 billion in 2008. Continued growth for the renewable energy sector is expected in the mid- to long-term, but 2009 will be a year of refocus, consolidation, or retrenchment for many companies. At the same time, new government spending, regulation, and policies should help the industry weather the current economic crisis better than many other sectors. Most notably, U.S. President Barack Obama's American Recovery and Reinvestment Act of 2009 includes more than $70 billion in direct spending and tax credits for clean energy and associated transportation programs. This policy-stimulus combination represents the largest federal commitment in U.S. history for renewables, advanced transportation, and energy conservation initiatives. Based on these new rules, many more utilities are expected to strengthen their clean-energy programs.

Clean Edge suggests that the commercialization of clean energy will help countries around the world pull out of the current economic malaise.

SUSTAINABLE ENERGY

Moving towards energy sustainability will require changes not only in the way energy is supplied, but in the way it is used, and reducing the amount of energy required to deliver various goods or services is essential. Opportunities for improvement on the demand side of the energy equation are as rich and diverse as those on the supply side, and often offer significant economic benefits.

Renewable energy and energy efficiency are said to be the "twin pillars" of sustainable energy policy. Any serious vision of a sustainable energy economy requires commitments to both renewables and efficiency. The American Council for an Energy-Efficient Economy has explained that both resources must be developed in order to stabilize and reduce carbon dioxide emissions:

Efficiency is essential to slowing the energy demand growth so that rising clean energy supplies can make deep cuts in fossil fuel use. If energy use grows too fast, renewable energy development will chase a receding target. Likewise, unless clean energy supplies come online rapidly, slowing demand growth will only begin to reduce total emissions; reducing the carbon content of energy sources is also needed.

The IEA has stated that renewable energy and energy efficiency policies should be viewed as complementary tools for the development of a sustainable energy future, instead of being developed in isolation.

ETHANOL FUEL

Ethanol fuel is ethanol (ethyl alcohol), the same type of alcohol found in alcoholic beverages. It can be used as a fuel, mainly as a biofuel alternative to gasoline, and is widely used in cars in Brazil. Because it is easy to manufacture and process and can be made from very common crops such as sugar cane and corn, it is an increasingly common alternative to gasoline in some parts of the world. This is a renewable resource (can be produced, unlike petroleum which cannot be produced and in time will be gone).

Anhydrous ethanol (ethanol with less than 1% water) can be blended with gasoline in varying quantities up to pure ethanol (E100), and most gasoline engines will operate well with mixtures of 10% ethanol (E10). Most cars on the road today in the U.S. can run on blends of up to 10% ethanol, and the use of 10% ethanol gasoline is mandated in some cities where harmful levels of auto emissions are possible.

Ethanol can be mass-produced by fermentation of sugar or by hydration of ethylene (ethene $CH_2=CH_2$) from petroleum and other sources. Current interest in ethanol mainly lies in bio-ethanol, produced from the starch or sugar in a wide variety of crops, but there has been considerable debate about how useful bio-ethanol will be in replacing fossil fuels in vehicles. Concerns relate to the large amount of arable land required for crops, as well as the energy and pollution balance of the whole cycle of ethanol production. Recent developments with cellulosic ethanol production and commercialization may allay some of these concerns.

According to the International Energy Agency, cellulosic ethanol could allow ethanol fuels to play a much bigger role in the future than previously thought. Cellulosic ethanol offers promise as resistant cellulose fibers, a major and universal component in plant cells walls, can be used to generate ethanol. Dedicated energy crops such as switchgrass are also promising cellulose sources that can be produced in many regions of the United States.

Chemistry

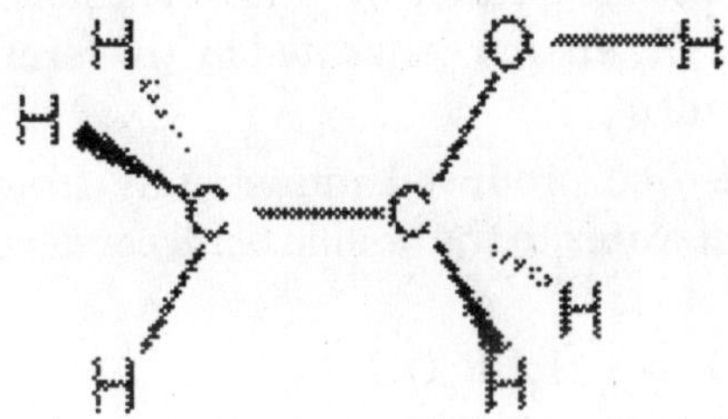

In this 3-d diagram of ethanol, the lines represent single bonds.

Glucose (a simple sugar) is created in the plant by photosynthesis.

$6CO_2 + 6H_2O + \text{light} \rightarrow C_6H_{12}O_6 + 6O_2$

During ethanol fermentation, glucose is decomposed into ethanol and carbon dioxide.

$C_6H_{12}O_6 \rightarrow 2C_2H_5OH + 2CO_2 + \text{heat}$

During combustion ethanol reacts with oxygen to produce carbon dioxide, water, and heat:

$C_2H_6O + 3O_2 \rightarrow 2CO_2 + 3H_2O + \text{heat}$

After doubling the ethanol combustion reaction because two molecules of ethanol are produced for each glucose molecule, there are equal numbers of each type of molecule on each side of the equation, and the net reaction for the overall production and consumption of ethanol is just:

$\text{light} \rightarrow \text{heat}$

The heat of the combustion of ethanol is used to drive the piston in the engine by expanding heated gases. It can be said that sunlight is used to run the engine. Glucose itself is not the only substance in the plant that is fermented. The simple sugar fructose also undergoes fermentation. Three other compounds in the plant can be fermented after breaking them up by hydrolysis into the glucose or fructose molecules that compose them. Starch and cellulose are molecules that are strings of glucose molecules, and sucrose (ordinary table sugar) is a molecule of glucose bonded to a molecule of fructose. The energy to create fructose in the plant ultimately comes from the metabolism of glucose created by photosynthesis, and so sunlight also provides the energy generated by the fermentation of these other molecules.

Ethanol may also be produced industrially from ethene (ethylene). Addition of water to the double bond converts ethene to ethanol:

$$CH_2{=}CH_2 + H_2O \rightarrow CH_3CH_2OH$$

This is done in the presence of an acid which catalyzes the reaction, but is not consumed. The ethene is produced from petroleum by steam cracking.

When ethanol is burned in the atmosphere rather than in pure oxygen, other chemical reactions occur with different components of the atmosphere such as N_2. This leads to the production of nitrous oxides NO_x , a major air pollutant.

Sources

Ethanol is a "renewable" because it is primarily the result of conversion of the sun's energy into usable energy. Creation of ethanol starts with photosynthesis causing the feedstocks such as switchgrass, sugar cane, or corn to grow. These feedstocks are processed into ethanol.

About 5% of the ethanol produced in the world in 2003 was actually a petroleum product. It is made by the catalytic hydration of ethylene with sulfuric acid as the catalyst. It can also be obtained via ethylene or acetylene, from calcium carbide, coal, oil gas, and other sources. Two million tons of petroleum-derived ethanol are produced annually. The principal suppliers are plants in the United States, Europe, and South Africa. Petroleum derived ethanol (synthetic ethanol) is chemically identical to bio-ethanol and can be differentiated only by radiocarbon dating.

Bio-ethanol is usually obtained from the conversion of carbon based feedstock. Agricultural feedstocks are considered renewable because they get energy from the sun using photosynthesis, provided that all minerals required for growth (such as nitrogen and phosphorus) are returned to the land. Ethanol can be produced from a variety of feedstocks such as sugar cane, bagasse, miscanthus, sugar beet, sorghum, grain sorghum, switchgrass, barley, hemp, kenaf, potatoes, sweet potatoes, cassava, sunflower, fruit, molasses, corn, stover, grain, wheat, straw, cotton, other biomass, as well as many types of cellulose waste and harvestings, whichever has the best well-to-wheel assessment.

An alternative process to produce bio-ethanol from algae is being developed by the company Algenol. Rather than grow algae and then harvest and ferment it the algae grow in sunlight and produce ethanol directly which is removed without killing the algae. It is claimed the process can produce 6000 gallons per acre per year compared with 400 gallons for corn production.

Currently, the first generation processes for the production of ethanol from corn use only a small part of the corn plant: the corn kernels are taken from the corn plant and only the starch, which represents about 50% of the dry kernel mass, is transformed into ethanol. Two types of second generation processes are under development. The first type uses enzymes and yeast to convert the plant cellulose into ethanol while the second type uses pyrolysis to convert the whole plant to either a liquid bio-oil or a syngas. Second generation processes can also be used with plants such as grasses, wood or agricultural waste material such as straw.

Production Process

The basic steps for large scale production of ethanol are: microbial (yeast) fermentation of sugars, distillation, dehydration (requirements vary, see Ethanol fuel mixtures, below), and denaturing (optional). Prior to fermentation, some crops require saccharification or hydrolysis of carbohydrates such as cellulose and starch into sugars. Saccharification of cellulose is called cellulolysis (see cellulosic ethanol). Enzymes are used to convert starch into sugar.

FERMENTATION

Ethanol is produced by microbial fermentation of the sugar. Microbial fermentation will currently only work directly with sugars. Two major components of plants, starch and cellulose, are both made up of sugars, and can in principle be converted to sugars for fermentation. Currently, only the sugar (e.g. sugar cane) and starch (e.g. corn) portions can be economically converted. However, there is much activity in the area of cellulosic ethanol, where the cellulose part of a plant is broken down to sugars and subsequently converted to ethanol.

Distillation

For the ethanol to be usable as a fuel, water must be removed. Most of the water is removed by distillation, but the purity is limited to 95-96% due to the formation of a low-boiling water-ethanol azeotrope. The 95.6% m/m (96.5% v/v) ethanol, 4.4% m/m (3.5% v/v) water mixture may be used as a fuel alone, but unlike anhydrous ethanol, is immiscible in gasoline, so the water fraction is typically removed in further treatment in order to burn with in combination with gasoline in gasoline engines.

Dehydration

There are basically five dehydration processes to remove the water from an azeotropic ethanol/water mixture. The first process, used in many early fuel ethanol plants, is called azeotropic distillation and consists of adding benzene or cyclohexane to the mixture. When these components are added to the mixture, it forms an heterogeneous azeotropic mixture in vapor-liquid-liquid equilibrium, which when distilled produces anhydrous ethanol in the column bottom, and a vapor mixture of water and cyclohexane/benzene. When condensed, this becomes a two-phase liquid mixture. Another early method, called extractive distillation, consists of adding a ternary component which will increase ethanol relative volatility. When the ternary mixture is distilled, it will produce anhydrous ethanol on the top stream of the column.

With increasing attention being paid to saving energy, many methods have been proposed that avoid distillation all together for dehydration. Of these methods, a third method has emerged and has been adopted by the majority of modern ethanol plants. This new process uses molecular sieves to remove water from fuel ethanol. In this process, ethanol vapor under pressure passes through a bed of molecular sieve beads. The bead's pores are sized to allow absorption of water while excluding ethanol. After a period of time, the bed is regenerated under vacuum to remove the absorbed water. Two beds are used so that one is available to absorb water while the other is being regenerated. This dehydration technology can account for energy saving of 3,000 btus/gallon compared to earlier azeotropic distillation. Modern Corn Ethanol plant description

Technology

Ethanol-based Engines

Ethanol is most commonly used to power automobiles, though it may be used to power other vehicles, such as farm tractors and airplanes. Ethanol (E100) consumption in an engine is approximately 51% higher than for gasoline since the energy per unit volume of ethanol is 34% lower than for gasoline. However, the higher compression ratios in an ethanol-only engine allow for increased power output and better fuel economy than could be obtained with lower compression ratios. In general, ethanol-only engines are tuned to give slightly better power and torque output than gasoline-powered engines. In flexible fuel vehicles, the lower compression ratio requires tunings that give the same output when using either gasoline or hydrated ethanol. For maximum use of ethanol's benefits, a much higher compression ratio should be used, which would render that engine unsuitable for gasoline use. When ethanol fuel availability allows high-compression ethanol-only vehicles to be practical, the fuel efficiency of such engines should be equal or greater than current gasoline engines. The mileage (miles-per-gallon) is therefore usually 20-30% higher than a gasoline-only engine.

A 2004 MIT study and an earlier paper published by the Society of Automotive Engineers identify a method to exploit the characteristics of fuel ethanol substantially better than mixing it with gasoline. The method presents the possibility of leveraging the use of alcohol to achieve definite improvement over the cost-effectiveness of hybrid electric. The improvement consists of using dual-fuel direct-injection of pure alcohol (or the azeotrope or E85) and gasoline, in any ratio up to 100% of either, in a turbocharged, high compression-ratio, small-displacement engine having performance similar to an engine having twice the displacement. Each fuel is carried separately, with a much smaller tank for alcohol. The high-compression (which increases efficiency) engine will run on ordinary gasoline under low-power cruise conditions. Alcohol is directly injected into the cylinders (and the gasoline injection simultaneously reduced) only when necessary to suppress 'knock' such as when significantly accelerating. Direct cylinder injection raises the already high octane rating of ethanol up to an effective 130. The calculated over-all reduction of gasoline use and CO_2 emission is 30%. The consumer cost payback time shows a 4:1 improvement over turbo-diesel and a 5:1 improvement over hybrid. In addition, the problems of water absorption into pre-mixed gasoline (causing phase separation), supply issues of multiple mix ratios and cold-weather starting are avoided.

In the UK one of the key criteria for determining the amount of required spinning reserve is the possible loss of "Sizewell B", a 1.32GW nuclear power plant.

At one point in the fall of 2007, out of 16 nuclear power stations in the UK, 7 were offline due to a combination of planned and unplanned outages.

Diesel Engine Generation

Small high speed diesels are very commonly used within large power grids throughout North America and Europe. France uses about 5 GW of such diesels to cover the intermittency of their nuclear stations; these are all in private hands - at small scales factories and the like - with their usage being triggered semi-randomly by a special tariff - EJP - which encourages these users to start their diesels. Large users such as CERN cut power usage as required by the System Operator - EDF.

In USA and UK these diesels have usually been purchased for other reasons e.g. for emergency standby, in water works, hotels, hospitals, etc. and in some cases for electricity substations - e.g. Cuyahoga Falls, USA (10 x 1.6 MW Caterpillar) and Tregarron Mid Wales UK (3 x 1.6 MW Caterpillar), but can be readily used to automatically synchronize and feed into the grid. In the UK 500 MW of such plant is routinely started within a few minutes; this is perfectly acceptable to the engines' service life in a scheme operated by National Grid called National Grid (UK) reserve service . It has been established that there is 20 GW of such diesel plant in the UK and it has been pointed out that there is no technical reason why this quantity could not be brought into the Reserve Service scheme to assist handling very rapid changes in renewable output, whilst conventional plant is started or indeed stopped.

Compensating for Variability

All sources of electrical power have some degree of unpredictability, and demand patterns routinely drive large swings in the amount of electricity that suppliers feed into the grid. Wherever possible, grid operations procedures are designed to match supply with demand at high levels of reliability, and the tools to influence supply and demand are well-developed. The introduction of large amounts of highly variable power generation may require changes to existing procedures and additional investments.

Operational Reserve

All managed grids already have existing operational and "spinning" reserve to compensate for existing uncertainties in the power grid. The addition of intermittent resources such as wind does not require 100% "back-up" because operating reserves and balancing requirements are calculated on a system-wide basis, and not dedicated to a specific generating plant.

- Some coal, gas, or hydro power plants are partially loaded and then controlled to change as demand changes or to replace rapidly lost generation. The ability to change as demand changes is termed "response." The ability to quickly replace lost generation, typically within timescales of 30 seconds to 30 minutes, is termed "spinning reserve."

- Generally thermal plants running as peaking plants will be less efficient than if they were running as base load.
- Hydroelectric facilities with storage capacity (such as the traditional dam configuration) may be operated as base load or peaking plants.
- In practice, as the power output from wind varies, partially loaded conventional plants, which are already present to provide response and reserve, adjust their output to compensate.
- While low penetrations of intermittent power may utilize existing levels of response and spinning reserve, the larger overall variations at higher penetrations levels will require additional reserves or other means of compensation.

Demand Reduction or Increase

- Demand response refers to the use of communication and switching devices which can release deferrable loads quickly, or absorb additional energy to correct supply/demand imbalances. Incentives have been widely created in the American, British and French systems for the use of these systems, such as favorable rates or capital cost assistance, encouraging consumers with large loads to take them off line or to start diesels whenever there is a shortage of capacity, or conversely to increase load when there is a surplus.
- Load Control allows the power company to turn loads off remotely if insufficient power is available.
- Energy demand management refers to incentives to adjust use of electricity, such as higher rates during peak hours.

- Real-time variable electricity pricing can encourage users to adjust usage to take advantage of periods when power is cheaply available and avoid periods when it is more scarce and expensive.
- Instantaneous demand reduction. Most large systems also have a category of loads which instantly disconnect when there is a generation shortage, under some mutually beneficial contract. This can give instant load reductions (or increases). See National Grid (UK) reserve service

Storage and Demand Loading

At times of low or falling demand where wind output may be high or increasing, grid stability may require lowering the output of various generating sources or even increasing demand, possibly by using energy storage to time-shift output to times of higher demand. Such mechanisms can include:

- Storage of electrical energy results in some lost energy because storage and retrieval are not perfectly efficient.
- Storage may require substantial capital investment and space for storage facilities.
- Pumped storage hydropower is the most prevalent existing technology used, and can substantially improve the economics of wind power. The availability of hydropower sites suitable for storage will vary from grid to grid. Typical round trip efficiency is 80%.
- Thermal energy storage stores heat or ice (or a similarly cold material). Stored heat can be used directly for heating needs or converted into electricity. Stored ice can be used as a source of air-conditioning during periods of high demand.
- Various other potential applications are being considered, such as charging plug-in electric vehicles during periods of low demand and high production; such technologies are not widely used at this time.
- Hydrogen can be created through electrolysis and stored for later use. NREL found that a kilogram of hydrogen

(roughly equivalent to a gallon of gasoline) could be produced for between $5.55 in the near term and $2.27 in the long term.

- Rechargeable flow batteries can serve as a large capacity, rapid-response storage medium.
- Some loads such as desalination plants and electric boilers, are able to store their output (water and heat.) These "opportunistic loads" are able to take advantage of "burst electricity" when it is available.

Geographic Diversity

The variability of production from a single wind turbine can be high. Combining any additional number of turbines (for example, in a wind farm) results in lower statistical variation, as long as the correlation between the output of each turbine is imperfect, and the correlations are always imperfect due to the distance between each turbine. Similarly, geographically distant wind turbines or wind farms have lower correlations, reducing overall variability. Since wind power is dependent on weather systems, there is a limit to the benefit of this geographic diversity for any power system. Multiple wind farms spread over a wide geographic area and gridded together produce power more constantly and with less variability than smaller installations. Wind output can be predicted with some degree of confidence using weather forecasts, especially from large numbers of turbines/farms. The ability to predict wind output is expected to increase over time as data is collected, especially from newer facilities.

Complementary Power Sources and Matching Demand

- Electricity produced from solar energy could be a counter balance to the fluctuating supplies generated from wind. In some locations, it tends to be windier at night and during cloudy or stormy weather, so there is likely to be more sunshine when there is less wind.
- In some locations, electricity demand may have a high correlation with wind output, particularly in locations where cold temperatures drive electric consumption (as cold air is denser and carries more energy).

- The allowable penetration may be further increased by increasing the amount of part-loaded generation available. Systems with existing high levels of hydroelectric generation may be able to incorporate substantial amounts of wind, although high hydro penetration may indicate that hydro is already a low-cost source of electricity; Norway, Quebec, and Manitoba all have high levels of existing hydroelectric generation (Quebec produces over 90% of its electricity from hydropower, and the local utility, Hydro-Québec, is the largest single hydropower producer in the world). The US Pacific Northwest has been identified as another region where wind energy is complemented well by existing hydropower, and there were "no fundamental technical barriers" to integrating up to 6000 MW of wind capacity. Storage capacity in hydropower facilities will be limited by size of reservoir, and environmental and other considerations.
- The Institute for Solar Energy Supply Technology of the University of Kassel pilot-tested a combined power plant linking solar, wind, biogas and hydrostorage to provide load-following power around the clock, entirely from renewable sources.

Export & Import Arrangements with Neighboring Systems

- It is often feasible to export energy to neighboring grids at times of surplus, and import energy when needed. This practice is common in Western Europe and North America.
- Integration with other grids can lower the effective concentration of variable power. Denmark's 44% penetration, in the context of the German/ Dutch/Scandinavian grids with which it has interconnections, is considerably lower as a proportion of the total system. This effect is diminished if more neighboring grids also have high penetration levels of variable power.

- Integration of grids may decrease the overall variability of both supply and demand by increasing geographical diversity.
- Methods of compensating for power variability in one grid, such as peaking-plants or pumped-storage hydro-electricity, may be taken advantage of by importing variable power from another grid that is short on such capabilities..
- The capacity of power transmission infrastructure may have to be substantially upgraded to support export/ import plans.
- Some energy is lost in transmission.
- The economic value of exporting variable power depends in part on the ability of the exporting grid to provide the importing grid with useful power at useful times for an attractive price.

Penetration

Penetration refers to the nominal capacity of a power source divided by peak demand in the system grid, expressed as a percentage. Penetration may also be used to refer to the amount of energy generated as a percentage of annual consumption.

- As penetration increases, the variations in power produced for the grid become larger and the costs and complexity of compensating for these variations increases.
- Large, geographically distributed networks may accept a higher penetration of wind than small networks because fluctuations in supply and demand across the entire grid can be averaged out.
- Power grids with significant amounts pumped storage, hydropower or other peaking power plants such as natural gas-fired power plants are more inherently capable of accommodating fluctuations from intermittent power.

- If an intermittent source produces more power than can be used, stored, or exported at that time, then that excess power will be lost.
- Wind power generation tends to be higher in the winter and at night (due to higher air density), so the appropriateness of wind power in high concentrations may crucially depend on the prevalence of air conditioning in a given jurisdiction. Wind power may be weakest in the hot summer months, and particularly during the day when air conditioning demand is highest. Conversely, systems where heat is electrical may be well-suited to higher penetration of wind power.
- Isolated, relatively small systems with only a few wind plants may only be stable and economic with a lower fraction of wind energy (e.g. Ireland), although mixed wind/diesel systems have been used in isolated communities with success at relatively high penetration levels.

The maximum proportion of intermittent power allowable in a power system will thus depend on many factors, including the size of the system, the attainable geographical diversity, the ability of the system to transmit power to where it is needed, storage capabilities, demand control capabilities, the conventional plant mix (coal, gas, nuclear, hydroelectric) and seasonal load factors (heating in winter, air-conditioning in summer) and their statistical correlation with power output.

There is no generally accepted maximum level of penetration, as each system's capacity to compensate for intermittency differs, and the systems themselves will change over time. Discussion of acceptable or unacceptable penetration figures should be treated and used with caution, as the relevance or significance will be highly dependent on local factors, grid structure and management, and existing generation capacity.

For most systems worldwide, existing penetration levels are significantly lower than practical or theoretical maximums; for example, a UK study found that "it is clear that intermittent generation need not compromise electricity system reliability at any level of penetration foreseeable in Britain over the next

20 years, although it may increase costs." As of 2006, Denmark has more than 40% penetration and at least two other countries (Portugal and Germany) have penetration levels (nominal to peak demand) of more than 20%.

MAXIMUM PENETRATION LIMITS

There is no generally accepted maximum penetration of wind energy that would be feasible in any given grid. Rather, economic efficiency and cost considerations are more likely to dominate as critical factors; technical solutions may allow higher penetration levels to be considered in future, particularly if cost considerations are secondary.

High penetration scenarios may be feasible in certain circumstances:

- Power generation for periods of little or no wind generation can be provided by retaining the existing power stations. The cost of using existing power stations for this purpose may be low since fuel costs dominate the operating costs. The actual cost of paying to keep a power station idle, but usable at short notice, may be estimated from published spark spreads and dark spreads. As existing traditional plant ages, the cost of replacing or refurbishing these facilities will become part of the cost of high-penetration wind if they are used only to provide operational reserve.
- Automatic load shedding of large industrial loads and its subsequent automatic reconnection is established technology and used in the UK and US, and known as Frequency Service contractors in the UK. Several GW are switched off and on each month in the UK in this way. Reserve Service contractors offer fast response gas turbines and even faster diesels in the UK, France and US to control grid stability.
- In a close-to-100% wind scenario, surplus wind power can be allowed for by increasing the levels of the existing Reserve and Frequency Service schemes and by extending the scheme to domestic-sized loads. Energy can be stored by advancing deferrable domestic loads

such as storage heaters, water heaters, fridge motors, or even hydrogen production, and load can be shed by turning such equipment off.

- Alternatively or additionally, power can be exported to neighboring grids and re-imported later. HVDC cables are efficient with 3% loss per 1000 km and may be inexpensive in certain circumstances. For example an 8 GW link from UK to France would cost about £1 billion using high-voltage direct current cables. Under such scenarios, the amount of transmission capacity required may be many times higher than currently available.

Penetration Studies

Studies have been conducted to assess the viability of specific penetration levels in specific energy markets.

European Supergrid

A series of detailed modelling studies by Dr. Gregor Czisch, which looked at the European wide adoption of renewable energy and interlinking power grids using HVDC cables, indicates that the entire European power usage could come from renewables, with 70% total energy from wind at the same sort of costs or lower than at present. This proposed large European power grid has been called a "super grid."

The model deals with intermittent power issues by using base-load renewables such as hydroelectric and biomass for a substantial portion of the remaining 30% and by heavy use of HVDC to shift power from windy areas to non-windy areas. The report states that "electricity transport proves to be one of the keys to an economical electricity supply" and underscores the importance of "international co-operation in the field of renewable energy use [and] transmission."

Dr. Czisch described the concept in an interview, saying "For example, if we look at wind energy in Europe. We have a winter wind region where the maximum production is in winter and in the Sahara region in northern Africa the highest wind production is in the summer and if you combine both, you come quite close to the needs of the people living in the whole area - let's say from northern Russia down to the southern part of the Sahara."

Grid Study in Ireland

A study of the grid in Ireland indicates that it would be feasible to accommodate 42% (of demand) renewables in the electricity mix. This acceptable level of renewable penetration was found in what the study called Scenario 5, provided 47% of electrical capacity (different from demand) with the following mix of renewable energies:

- 6000MW wind
- 360MW base load renewables
- 285MW additional variable renewables (other intermittent sources)

The study cautions that various assumptions were made that "may have understated dispatch restrictions, resulting in an underestimation of operational costs, required wind curtailment, and CO2 emissions" and that "The limitations of the study may overstate the technical feasability of the portfolios analysed..."

Scenario 6, which proposed renewables providing 59% of electrical capacity and 54% of demand had problems. Scenario 6 proposed the following mix of renewable energies:

- 8000MW wind
- 392MW base load renewables
- 1685MW additional variable renewables (other intermittent sources)

The study found that for Scenario 6, "a significant number of hours characterised by extreme system situations occurred where load and reserve requirements could not be met. The results of the network study indicated that for such extreme renewable penetration scenarios, a system re-design is required, rather than a reinforcement excercise." The study declined to analyze the cost effectiveness of the required changes because "determination of costs and benefits had become extremely dependent on the assumptions made" and this uncertainty would have impacted the robustness of the results.

Ontario

A study published in October, 2006, by the Ontario Independent Electric System Operator (IESO) found that "there would be minimal system operation impacts for levels of wind capacity up to 5,000 megawatts (MW)," which corresponds to a peak penetration of 17%.

ECONOMIC IMPACTS OF VARIABILITY

Estimates of the cost of wind energy may include estimates of the "external" costs of wind variability, or be limited to the cost of production. All electrical plant has costs that are separate from the cost of production, including, for example, the cost of any necessary transmission capacity or reserve capacity in case of loss of generating capacity. Many types of generation, particularly fossil fuel derived, will also have cost externalities such as pollution, greenhouse gas emission, and habitat destruction which are generally not directly accounted for. The magnitude of the economic impacts is debated and will vary by location, but is expected to rise with higher penetration levels. At low penetration levels, costs such as operating reserve and balancing costs are believed to be insignificant.

Intermittency may introduce additional costs that are distinct from or of a different magnitude than for traditional generation types. These may include:

- Transmission capacity: transmission capacity may be more expensive than for nuclear and coal generating capacity due to lower load factors. Transmission capacity will generally be sized to projected peak output, but average capacity for wind will be significantly lower, raising cost per unit of energy actually transmitted.
- Additional operating reserve: if additional wind does not correspond to demand patterns, additional operating reserve may be required compared to other generating types, resulting in higher capital costs for additional plants. Contrary to statements that all wind must be backed by an equal amount of "back-up capacity", intermittent generators contribute to base capacity "as long as there is some probability of output during peak

periods." Back-up capacity is not attributed to individual generators, as back-up or operating reserve "only have meaning at the system level."

- Balancing costs: to maintain grid stability, some additional costs may be incurred for balancing of load with demand. The ability of the grid to balance supply with demand will depend on the rate of change of the amount of energy produced (by wind, for example) and the ability of other sources to ramp production up or scale production down. Balancing costs have generally been found to be low.
- Storage, export and load management: at high penetrations (more than 30%), solutions (described below) for dealing with high output of wind during periods of low demand may be required. These may require additional capital expenditures, or result in lower marginal income for wind producers.

Analyses of Costs

Studies have been performed to determine the costs of variability. The British Wind Energy Association states:

"A review of integration studies, worldwide, suggests that the additional costs of integrating wind are around £2/MWh with 10% wind, rising to £3/MWh with 20% wind."

Colorado - Separate reports by Xcel and UCS

An official at Xcel Energy claimed that at 20 percent penetration, additional standby generators to compensate for wind in Colorado would cost $8 per MWh, adding between 13% and 16% to the $50-$60 cost per MWh of wind energy.

The Union of Concerned Scientists conducted a study of the costs to increase the renewable penetration in Colorado to 10% and found that for an average residential bill "customers of municipal utilities and rural electric cooperatives that opt out of the solar energy requirement" would save 4 cents per month, but that for Xcel Energy customers there would be additional cost of about 10 cents per month. Total impact on all consumers would be $4.5 million or 0.01% over two decades.

UK Studies

National Grid study of 10% renewables by 2010

A detailed study for UK National Grid (a private power company) states "We have estimated that for the case with 8000MW of wind needed to meet the 10% renewables target for 2010, balancing costs can be expected to increase by around £2 per MWh of wind production. This would represent an additional £40million per annum, just over 10% of existing annual balancing costs."

National Grid study of 40% renewables

In evidence to the UK House of Lords Economic Affairs Select Committee, National Grid have quoted estimates of balancing costs for 40% wind and these lie in the range £500-1000M per annum. "These balancing costs represent an additional £6 to £12 per annum on average consumer electricity bill of around £390." National Grid notes that "increasing levels of such renewable generation on the system would increase the costs of balancing the system and managing system frequency."

Carbon Trust & DTI study of 10% by 2010

A 2003 report by Carbon Trust and the UK Department of Trade and Industry (DTI) found that the costs for reinforcement and new build of transmission and distribution systems to support 10% renewable electricity in the UK by 2010 would be £1.6billion to £2.4bn. The study classified "Intermittency" as "Not a significant issue" for the 2010 target.

Carbon Trust & DTI study of 20% by 2020

The same 2003 study found that achieving 20% renewable electricity in the UK by 2020 would cost £3.2bn to £4.5bn in transmission and distribution system construction and reinforcement. The study classified "Intermittency" as a "Significant Issue" for the 2020 target.

Minnesota

Minnesota study on wind penetration levels and found that "total integration operating cost for up to 25% wind energy" would be less than $0.0045 per kWh (additional).

INTERMITTENCY AND RENEWABLE ENERGY

There are differing views about renewable energy and intermittency. The World Nuclear Association argues that the sun, wind, tides and waves cannot be controlled to provide directly either continuous base-load power, or peak-load power when it is needed. Proponents of renewable energy use argue that the issue of intermittency of renewables is over-stated, and that practical experience demonstrates this.

Views of Critics of High Penetration Renewable Energy Use

The World Nuclear Association states that:

> *"Sun, wind, tides and waves cannot be controlled to provide directly either continuous base-load power, or peak-load power when it is needed. In practical terms they are therefore limited to some **10-20%** of the capacity of an electricity grid, and cannot directly be applied as economic substitutes for coal or nuclear power, however important they may become in particular areas with favourable conditions."*
>
> *"The fundamental problem, especially for electricity supply, is their variable and diffuse nature. This means either that there must be reliable duplicate sources of electricity, or some means of electricity storage on a large scale. Apart from pumped-storage hydro systems, no such means exist at present and nor are any in sight."*

"Relatively few places have scope for pumped storage dams close to where the power is needed, and overall efficiency is low. Means of storing large amounts of electricity as such in giant batteries or by other means have not been developed."

On December 10, 2007 Patrick Moore, co-chai of the Clean & Safe Energy Coalition - a pro-nuclear group funded by the Nuclear Energy Institute - wrote:

> *"Greenpeace is deliberately misleading the public into thinking that wind and solar energy, both of which are inherently intermittent and unreliable, can replace*

*baseload power that is continuous and reliable. Only three technologies can produce large amounts of baseload power: fossil fuels, hydroelectric plants and nuclear power. Given that we want to reduce fossil fuels and that potential hydroelectric sites are becoming scarce, nuclear power is the main option... Over the past **10** years, Germany and Denmark have poured billions of taxpayers' euros into wind and solar energy in the vain hope that this would allow them to shut down fossil fuel and nuclear plants. They have not succeeded because every solar panel and every wind turbine must be backed up by reliable power when the sun isn't shining and the wind isn't blowing."*

Mr. Moore is a co-founder and former leader of Greenpeace, but he has not been involved with Greenpeace since 1986.

Views of Proponents of High Penetration Renewable Energy Use

Some renewable electricity sources have identical variability to coal-fired power stations, so they are base-load, and can be integrated into the electricity supply system without any additional back-up. Examples include:

- Bio-energy, based on the combustion of crops and crop residues, or their gasification followed by combustion of the gas.
- Hot dry rock geothermal power, which is being developed in Australia and the United States.
- Solar thermal electricity, with overnight heat storage in water or rocks, or a thermochemical store as with Nevada Solar One and Solar Tres.

Furthermore, supporters argue that the total electricity generated from a large-scale array of dispersed wind farms, located in different wind regimes, cannot be accurately described as intermittent, because it does not start up or switch off instantaneously at irregular intervals. With a small amount of supplementary peak-load plant, which operates infrequently, large-scale distributed wind power can substitute for some base-load power and be equally reliable.

Hydropower can be intermittent and/or dispatchable, depending on the configuration of physical plant. Typical hydroelectric plants in the dam configuration may have substantial storage capacity, and be considered dispatchable. Run of the river hydroelectric generation will typically have limited or no storage capacity, and will be variable on a seasonal or annual basis (dependent on rainfall and snow melt).

Amory Lovins suggests a few basic strategies to deal with these issues:

> *"The variability of sun, wind and so on, turns out to be a non-problem if you do several sensible things. One is to diversify your renewables by technology, so that weather conditions bad for one kind are good for another. Second, you diversify by site so they're not all subject to the same weather pattern at the same time because they're in the same place. Third, you use standard weather forecasting techniques to forecast wind, sun and rain, and of course hydro operators do this right now. Fourth, you integrate all your resources — supply side and demand side..."*

Moreover, efficient energy use and energy conservation measures can reliably reduce demand for base-load and peak-load electricity.

Several studies have demonstrated the technical feasibility of integrating intermittent power at levels substantially higher than is common in most countries (from 15-30% penetration), and at least three countries have more than 20% wind penetration. Relatively few changes to large grids are normally required and the associated system costs are moderate. International groups are studying much higher penetrations (30-75%, corresponding to up to 20% of national electricity consumption) and preliminary conclusions are that these levels are also technically feasible. In the UK, one summary of other studies indicated that if assuming that wind power contributed less than 20% of UK power consumption, then the intermittency would cause only moderate cost.

Methods to manage wind power integration range from those that are commonly used at present (e.g. demand management) to potential new technologies for grid energy storage. Improved forecasting can also contribute as the daily and seasonal variations in wind and solar sources are to some extent predictable.

The Pembina Institute and the World Wide Fund for Nature state in the Renewable is Doable plan that resilience is a feature of renewable energy:

> *"Diversity and dispersal also add system security. If one wind turbine fails, the lights won't flicker. If an entire windfarm gets knocked out by a storm, only 40,000 people will lose power. If a single Darlington reactor goes down, 400,000 homes, or key industries, could face instant blackouts. To hedge this extra risk, high premiums have to be paid for decades to ensure large blocks of standby generation.*

ETHANOL FUEL ENERGY BALANCE

All biomass needs to go through some of these steps: it needs to be grown, collected, dried, fermented, and burned. All of these steps require resources and an infrastructure. The total amount of energy input into the process compared to the energy released by burning the resulting ethanol fuel is known as the ethanol fuel energy balance (sometimes called "Net energy gain") and studied as part of the wider field of energy economics. Figures compiled in a 2007 National Geographic Magazine article point to modest results for corn ethanol produced in the US: 1 unit of current energy equals 1.3 energy units of corn ethanol energy. The energy balance for sugarcane ethanol produced in Brazil is more favorable, 1:8. Over the years, however, many reports have been produced with contradicting energy balance estimates.

ENERGY BALANCE REPORTS

In 1995 the USDA released a report stating this is that the net energy balance of corn ethanol in the United States was an average of 1.24. It was previously considered to have a negative net energy balance. However, due to increases in corn crop yield and more efficient farming practices corn ethanol had gained energy efficiency. Cassman, a Heuermann Professor of Agronomy at the university, said ethanol has a substantial net positive direct energy balance — 1.5 to 1.6 more units of energy are derived from ethanol than are used to produce it. Compared to just five years ago, Tiemann said, ethanol plants produce 15 percent more ethanol from a bushel of corn and use about 20 percent less energy in the process. At the same time, corn growers are more efficient, producing more corn per acre and using less energy to do so.

Opponents of corn ethanol production in the U.S. often quote the 2005 paper of David Pimentel, a retired Entomologist, and Tadeusz Patzek, a Geological Engineer from UC Berkeley. Both have been exceptionally critical of ethanol and other biofuels. Their studies contend that ethanol, and biofuels in general, are "energy negative", meaning they take more energy to produce than is contained in the final product.

A 2006 article in Science offers the consensus opinion that current corn ethanol technologies had similar greenhouse gas emissions to gasoline, but was much less petroleum-intensive than gasoline. Furthermore, it should be pointed out that fossil fuels also require significant energy inputs which have seldom been accounted for in the past.

It is also important to note that ethanol is not the only product created during production, and the energy content of the by-products must also be considered. Corn is typically 66% starch and the remaining 33% is not fermented. This unfermented component is called distillers grain, which is high in fats and proteins, and makes good animal feed.

Back in 2000, Dr. Michael Wang, of Argonne National Laboratory, wrote that these ethanol by-products are the most contentious issue in evaluating the energy balance of ethanol. He wrote that Pimentel assumes that corn ethanol entirely replaces gasoline and so the quantity of by-products is too large for the market to absorb, and they become waste. At lower quantities of production, Wang finds it appropriate to credit corn ethanol based on the input energy requirement of the feed product or good that the ethanol by-product displaces. In 2004, a USDA report found that co-products accounting made the difference between energy ratios of 1.06 and 1.67. In 2006, MIT researcher Tiffany Groode came to similar conclusions about the co-product issue.

In Brazil where sugar cane is used, the yield is higher, and conversion to ethanol is somewhat more energy efficient than corn. Recent developments with cellulosic ethanol production may improve yields even further.

In 2006 a study from the University of Minnesota found that corn-grain ethanol produced 1.25 units of energy per unit put in.

A 2008 study by the University of Nebraska found a 5.4 energy balance for ethanol derived specifically from switchgrass. This estimate is better than in previous studies and according to the authors partly due to the larger size of the field trial (3-9 ha) on 10 farms.

Variables

According to DoE, to evaluate the net energy of ethanol four variables must be considered:

1. the amount of energy contained in the final ethanol product
2. the amount of energy directly consumed to make the ethanol (such as the diesel used in tractors)
3. the quality of the resulting ethanol compared to the quality of refined gasoline
4. the energy indirectly consumed (in order to make the ethanol processing plant, etc).

Much of the current academic discussion regarding ethanol currently revolves around issues of system borders. This refers to how complete of a picture is drawn for energy inputs. There is debate on whether to include items like the energy required to feed the people tending and processing the corn, to erect and repair farm fences, even the amount of energy a tractor represents.

In addition, there is no consensus on what sort of value to give the rest of the corn (such as the stalk), commonly known as the 'coproduct.' Some studies leave it on the field to protect the soil from erosion and to add organic matter, while others take and burn the coproduct to power the ethanol plant, but do not address the resulting soil erosion (which would require energy in the form of fertilizer to replace). Depending on the ethanol study you read, net energy returns vary from .7-1.5 units of ethanol per unit of fossil fuel energy consumed. For comparison, that same one unit of fossil fuel invested in oil and gas extraction (in the lower 48 States) will yield 15 units of gasoline, a yield an order of magnitude better than current ethanol production technologies, ignoring the energy quality arguments above and the fact that the gain (14 units) is not carbon neutral.

In this regard, geography is the decisive factor. In tropical regions with abundant water and land resources, such as Brazil and Colombia, the viability of production of ethanol from sugarcane is no longer in question; in fact, the burning of sugarcane residues (bagasse) generates far more energy than needed to operate the ethanol plants, and many of them are now selling electric energy to the utilities. However, while there may be a positive net energy return at the moment, recent research suggests that the sugarcane plantations are not sustainable in the long run, as they are depleting the soil of nutrients and carbon matter. Did they return the ashes from their plants to the land in the recent research?

The picture is different for other regions, such as most of the United States, where the climate is too cool for sugarcane. In the U.S., agricultural ethanol is generally obtained from grain, chiefly corn. But it can also be obtained from cellulose, more energy balanced bioethanol.

Clean Production Bioethanol

Clean production bioethanol is a biofuel obtained using as much as possible non-greenhouse gas renewable energy sources:

- The amount of energy directly consumed to make the ethanol is renewable energy : the tractor uses ethanol engine, biodiesel, air engine or electricity obtained from wind or solar energy.
- The energy indirectly consumed, that can be solar energy.

To transport the biofuel to the fuel-stations can be used truck with ethanol engines or electric motors (that uses energy solar energy stored in the batteries).

CELLULOSIC ETHANOL COMMERCIALIZATION

Cellulosic ethanol commercialization is the process of building an industry out of methods of turning cellulose-containing organic matter into fuel. Companies such as Iogen, Broin, and Abengoa are building refineries that can process biomass and turn it into ethanol, while companies such as Diversa, Novozymes, and Dyadic are producing enzymes which could enable a cellulosic ethanol future. The shift from food crop feedstocks to waste residues and native grasses offers significant opportunities for a range of players, from farmers to biotechnology firms, and from project developers to investors.

CELLULOSIC ETHANOL PRODUCTION

Cellulosic ethanol can be produced from a diverse array of feedstocks. Instead of taking the grain from wheat and grinding that down to get starch and gluten, then taking the starch, cellulosic ethanol production involves the use of the whole crop. This approach should increase yields and reduce the carbon footprint because the amount of energy-intensive fertilisers and fungicides will remain the same, for a higher output of usable material.

COMMERCIALIZATION BY COUNTRY

Australia

Ethtec has a pilot plant in Harwood, New South Wales, which uses wood residues as a feedstock.

Canada

In Canada, Iogen Corp. is a developer of cellulosic ethanol process technology. Iogen has developed a proprietary process and operates a demonstration-scale plant in Ontario. The facility has been designed and engineered to process 40 tons of wheat straw per day into ethanol using enzymes made in an adjacent enzyme manufacturing facility. In 2004, Iogen began delivering its first shipments of cellulosic ethanol into the marketplace. In the near term, the company intends to commercialize its cellulose ethanol process by licensing its technology broadly through turnkey plant construction partnerships. The company is currently evaluating sites in the United States and Canada for its first commercial-scale plant.

Lignol Innovations has a pilot plant, which uses wood as a feedstock, in Vancouver.

In March 2009, KL Energy Corporation of South Dakota and Prairie Green Renewable Energy of Alberta announced their intention to develop a cellulosic ethanol plant near Hudson Bay, Saskatchewan. The Northeast Saskatchewan Renewable Energy Facility will use KL Energy's modern design and engineering to produce ethanol from wood waste.

China

Cellulosic ethanol production currently exists at "pilot" and "commercial demonstration" scale, including a plant in China engineered by SunOpta Inc. and owned and operated by China Resources Alcohol Corporation that is currently producing cellulosic ethanol from corn stover (stalks and leaves) on a continuous, 24-hour per day basis.

Denmark

Dong Energy has a pilot plant in Kalundborg, which uses wheat straw as a feedstock.

Japan

Nippon Oil Corporation and other Japanese manufacturers including Toyota Motor Corporation plan to set up a research body to develop cellulose-derived biofuels. The consortium plans to produce 250,000 kilolitres (1.6 million barrels) per year of bioethanol by March 2014, and produce bioethanol at 40 yen ($0.437) per litre (about $70 a barrel) by 2015.

In March 2009, Honda Motor announced an agreement for the construction of a new cellulosic ethanol research facility in Japan. The new Kazusa-branch facility of the Honda Fundamental Technology Research Center will be built within the Kazusa Akademia Park, in Kisarazu, Chiba. Construction is scheduled to begin in April 2009, with the aim to begin operations in November 2009.

Spain

Abengoa continues to invest heavily in the necessary technology for bringing cellulosic ethanol to market. Utilizing process and pre-treatment technology from SunOpta Inc., Abengoa is building a 5 million gallon cellulosic ethanol facility in Spain and have recently entered into a strategic research and development agreement with Dyadic International, Inc. (AMEX: DIL), to create new and better enzyme mixtures which may be used to improve both the efficiencies and cost structure of producing cellulosic ethanol.

Sweden

SEKAB has developed an industrial process for production of ethanol from biomass feed-stocks, including wood chips and sugar cane bagasse. The development work is being carried out at an advanced pilot plant in Örnsköldsvik, and has sparked international interest. The technology will be gradually scaled up to commercial production in a new breed of bio-refineries from 2013 to 2015.

United Kingdom

A $400 million investment programme to cover the construction of a world scale ethanol plant and a high technology demonstration plant to advance development work on the next generation of biofuels has been announced by BP, Associated British Foods (ABF) and DuPont. The bioethanol plant will be built on BP's existing chemicals site at Saltend, Hull. Due to be commissioned in late 2009, it will have an annual production capacity of some 420 million litres from wheat feedstock.

United States

The U.S. Department of Energy (DOE) is promoting the development of ethanol from cellulosic feedstocks as an alternative to conventional petroleum transportation fuels. Programs sponsored by DOE range from research to develop better cellulose hydrolysis enzymes and ethanol-fermenting organisms, to engineering studies of potential processes, to co-funding initial ethanol from cellulosic biomass demonstration and production facilities. This research is conducted by various national laboratories, including the National Renewable Energy Laboratory (NREL), Oak Ridge National Laboratory (ORNL) and Idaho National Laboratory (INL), as well as by universities and private industry. Engineering and construction companies and operating companies are generally conducting the engineering work.

In May 2008, Congress passed a new farm bill that will accelerate the commercialization of advanced biofuels, including cellulosic ethanol. The *Food, Conservation, and Energy Act of 2008* provides for grants covering up to 30% of the cost of developing and building demonstration-scale biorefineries for producing "advanced biofuels," which essentially includes all fuels that are not produced from corn kernel starch. It also allows for loan guarantees of up to $250 million for building commercial-scale biorefineries to produce advanced biofuels.

In September 2008, Iogen Corporation delivered more than 100,000 liters (26,000 gallons) of cellulosic ethanol to Royal Dutch Shell, its commercial business partner, "for upcoming fuel applications." The fuel was the first part of Shell's initial order of cellulosic ethanol from Iogen, which totaled 180,000 liters (47,550 gallons). Shell first gained an equity stake in Iogen in 2002, and in July 2008, Shell increased its shareholding in Iogen from slightly more than 26 percent to 50 percent.

Following a successful start-up in the fourth quarter of 2008, the POET LLC Research Center in Scotland, South Dakota is now producing cellulosic ethanol on a pilot scale, at a rate of 20,000 gallons-per-year (GPY) using corn cobs as feedstock. The US $8 million plant is a precursor to the company's US $200 million Project Liberty, a commercial-scale cellulosic ethanol plant which is to begin production in 2011.

Verenium Corporation announced in January 2009 that it will construct a commercial-scale cellulosic ethanol plant in Highlands County, Florida. The plant, which will provide about 150 full-time jobs in the production of 36 million gallons of ethanol per year, will cost about $250 million to construct. Verenium plans to break ground in the second half of 2009 and have the plant in full production during 2011.

Commercial Cellulosic Ethanol Plants in the U.S. (Operational or under construction)		
Company	**Location**	**Feedstock**
Abengoa Bioenergy	Hugoton, KS	Wheat straw
BlueFire Ethanol	Irvine, CA	Multiple sources
Colusa Biomass Energy Corporation	Sacramento, CA	Waste rice straw
Fulcrum BioEnergy	Reno, NV	Municipal solid waste
Gulf Coast Energy	Mossy Head, FL	Wood waste
KL Energy Corp.	Upton, WY	Wood
Mascoma	Lansing, MI	Wood
POET LLC	Emmetsburg, IA	Corn cobs
Range Fuels	Treutlen County, GA	Wood waste
SunOpta	Little Falls, MN	Wood chips
US Envirofuels	Highlands County, FL	Sweet sorghum
Xethanol	Auburndale, FL	Citrus peels

Using a newly developed tool known as the "Biofuels Deployment Model", Sandia researchers have determined that 21 billion gallons of cellulosic ethanol could be produced per year by 2022 without displacing current crops. The Renewable Fuels Standard, part of the 2007 Energy Independence and Security Act, calls for an increase in biofuels production to 36 billion gallons a year by 2022.

Environmental Issues

Cellulosic ethanol and grain-based ethanol are, in fact, the same product, but many scientists believe cellulosic ethanol production has distinct environmental advantages over grain-based ethanol production. On a life-cycle basis, ethanol produced from agricultural residues or dedicated cellulosic crops has significantly lower greenhouse gas emissions and a higher sustainability rating than ethanol produced from grain. According to US Department of Energy studies conducted by the Argonne Laboratories of the University of Chicago, cellulosic ethanol reduces greenhouse gas emissions (GHG) by 85% over reformulated gasoline. By contrast, starch ethanol (e.g., from corn), which usually uses natural gas to provide energy for the process, reduces greenhouse gas emissions by 18% to 29% over gasoline.

Criticism

Critics such as Cornell University professor of ecology and agriculture David Pimentel and University of California at Berkeley engineer Tad Patzek question the likelihood of environmental, energy, or economic benefits from cellulosic ethanol technology from non-waste.

ENERGY SECURITY AND RENEWABLE TECHNOLOGY

The environmental benefits of renewable energy technologies are widely recognised, but the contribution that they can make to energy security is less well known. Renewable technologies can enhance energy security in electricity generation, heat supply, and transportation.

Energy Security

Access to cheap energy has become essential to the functioning of modern economies. However, the uneven distribution of fossil fuel supplies among countries, and the critical need to widely access energy resources, has led to significant vulnerabilities. Threats to global energy security include political instability of energy producing countries, manipulation of energy supplies, competition over energy sources, attacks on supply infrastructure, as well as accidents and natural disasters.

Transportation

The International Energy Agency's *World Energy Outlook 2006* concludes that rising oil demand, if left unchecked, would accentuate the consuming countries' vulnerability to a severe supply disruption and resulting price shock. Renewable biofuels for transport represent a key source of diversification from petroleum products. Biofuels from grain and beet in temperate regions have a part to play, but they are relatively expensive and their energy efficiency and CO_2 savings benefits, are variable. Biofuels from sugar cane and other highly productive tropical crops are much more competitive and beneficial. But all first generation biofuels ultimately compete with food production for land, water, and other resources. Greater efforts are required to develop and commercialize second generation biofuel technologies, such as biorefineries and ligno-cellulosics, enabling the flexible production of biofuels and other products from non-edible plant materials.

According to the International Energy Agency (IEA), cellulosic ethanol commercialization could allow ethanol fuels to play a much larger role in the future than previously thought. Cellulosic ethanol can be made from plant matter composed primarily of inedible cellulose fibers that form the stems and branches of most plants. Dedicated energy crops, such as switchgrass, are also promising cellulose sources that can be produced in many regions of the United States.

Heating

For those countries where growing dependence on imported gas is a significant energy security issue, renewable technologies can provide alternative sources of electric power as well as displacing electricity demand through direct heat production. The IEA suggests that the direct contribution that renewables can make to domestic or commercial space heating and industrial process heat deserves closer attention. Heat from solar, and geothermal sources, as well as heat pumps, is increasingly cost effective but often falls through the gap between government programmes that promote public awareness and provide incentives for renewable electricity and energy efficiency.

Solar heating systems are a well known technology and generally consist of solar thermal collectors, a fluid system to move the heat from the collector to its point of usage, and a reservoir or tank for heat storage. The systems may be used to heat domestic hot water, swimming pools, or homes and businesses. The heat can also be used for industrial process applications or as an energy input for other uses such as cooling equipment. In many warmer climates, a solar heating system can provide a very high percentage (50 to 75%) of domestic hot water energy.

Electricity Generation

The deployment of renewable technologies usually increases the diversity of electricity sources and, through local generation, contributes to the flexibility of the system and its resistance to central shocks. The IEA suggests that attention in this area has focused disproportionately on the issue of the variability of renewable electricity production. However, this only applies to certain renewable technologies, mainly wind power and solar photovoltaics, and its significance depends on a range of factors which include the penetration of the renewables concerned, the balance of plant on the system, the wider connectivity of the system, and the demand side flexibility. Variability will rarely be a barrier to increased renewable energy deployment. But at high levels of penetration it requires careful analysis and management, and any additional costs that may be required for back-up or system modification must be taken into account.

COMBINED POWER PLANT

The Combined Power Plant, a project linking 36 wind, solar, biomass, and hydroelectric installations throughout Germany, has demonstrated that a combination of renewable sources and more-effective control can balance out short-term power fluctuations and provide reliable electricity with 100 percent renewable energy.

7

Renewable Energy Development

Renewable energy development covers the advancement, capacity growth, and use of renewable energy sources. Modern interest in renewable energy development is linked to concerns about exhaustion and greenhouse gases of fossil fuels and environmental, social and political risks of extensive use of fossil fuels and nuclear energy. It is a form of energy development with a focus on renewable energy, including all energy from renewable sources.

History

What is now thought of as renewable energy has been used by man since prehistory. The burning of biomass for heat and light has been practiced throughout recorded history. Windmills and watermills have converted the potential energy of water for centuries, to a power source for small-scale agricultural and industrial processes.

The modern technologies which now comprise renewables have different and varied histories. The beginning of the development of wind technology can be dated back to the late 19th Century and experiments in Denmark and elsewhere. Interest in the technology peaked in Denmark during both World Wars due to limited access to fossil fuels. From the 1950s onwards, photovoltaic (solar) cells saw investment as a result of their usefulness in space craft, with resulting improvements in the technology and knowledge of materials, along with reductions in price to levels acceptable to some consumers. The main motivation for the expansion of renewable energies came with the oil crises of 1973 and 1979-80. Concern by political leaders in a host of countries saw increases in support for research and development of new technologies (Jimmy Carter was the first US politician to focus significantly on solar energy use, in response to the 1973 crisis). Wind, wave and solar energy technologies all benefited from this investment with an increase in the range of their application.

Elsewhere, there have been experimentations with passive solar energy, including daylighting, which continues a tradition of orienting house-building to benefit from natural resources. Alongside the industrial development of renewable energies, many societies simply continue to practice as a matter of course what has happened for hundreds of years.

NON-TECHNICAL BARRIERS TO ACCEPTANCE

In September 2006, the U.S. Department of Energy's Office of Energy Efficiency and Renewable Energy did a review of recent literature discussing the "non-technical barriers" to renewable energy use, specifically solar. These are marketing, institutional, and policy impediments which are holding back the acceptance of renewable energy technologies. These key barriers are listed here, from most frequently cited to least, and "must be addressed as part of the technology acceptance efforts":

- Lack of government policy support. This includes the lack of policies and regulations supporting development of renewable energy technologies and the presence of policies and regulations hindering renewable energy development and supporting conventional energy development. Examples include fossil-fuel subsidies, insufficient consumer-based renewable energy incentives, government underwriting for nuclear plant accidents, and difficult zoning and permitting processes for renewable energy.
- Lack of information dissemination and consumer awareness.
- High capital cost of renewable energy technologies compared with conventional energy.
- Difficulty overcoming established energy systems. This includes difficulty introducing innovative energy

systems, particularly for distributed generation such as photovoltaics, because of technological lock-in, electricity markets designed for centralized power plants, and market control by established generators.

- Inadequate financing options for renewable energy projects.
- Failure to account for all costs and benefits of energy choices. This includes failure to internalize all costs of conventional energy (e.g., effects of air pollution, risk of supply disruption) and failure to internalize all benefits of renewable energy (e.g., cleaner air, energy security).
- Inadequate workforce skills and training. This includes lack in the workforce of adequate scientific, technical, and manufacturing skills required for renewable energy development; lack of reliable installation, maintenance, and inspection services; and failure of the educational system to provide adequate training in new technologies.
- Lack of adequate codes, standards, and interconnection and net-metering guidelines.
- Poor perception by public of renewable energy system aesthetics.
- Lack of stakeholder/community participation in energy choices and renewable energy projects.

RENEWABLE ENERGY SUPPORT MECHANISMS

Because of the above non-technical barriers, creating/ harvesting (renewable) energy may be more difficult and expensive than burning fossil fuels. Technology for using fossil fuels, such as mines and power plants, is already well established whereas much renewable technology is new. There remains considerable discussion amongst politicians, academics and others as to the most appropriate mechanism - or combination of mechanisms - for achieving renewable energy policy goals.

Some people claim that renewable energy is not *cost effective*, as it often needs government incentives in order to be viable. This may be because fossil fuel prices do not include the true costs of global ecological change now and in the future;

an argument of ecological economists. It should also be noted that the effort involved in extracting oil from ever deeper reservoirs is increasing and that the costs of renewable energy technology have been shown to fall with increased investment and expansion of capacity. Conceivably, the cost of renewable energy will drop below fossil fuel costs.

Much of renewable energy policy concerns the stimulation of markets and thus demand for the various technologies with the aim of improving efficiency and reducing costs. A number of policy mechanisms have been applied to this end, the mechanisms which have been applied widely are the quota (or RPS) mechanism, the fixed tariff mechanism, contract bidding mechanisms and the application of tax credits. Growing levels of experience in the application of these has enabled more critical assessments of their usefulness in addressing the multiple goals of renewable energy policy.

Quota Mechanisms

A quota mechanism, sometimes known as a Renewable Portfolio Standard (RPS), sees a government place an obligation on either an electricity supply company or on consumers (albeit usually manifested through their supply company) to source a specified fraction of their electricity from renewable energy sources. Companies which fail to meet the obligation are required to pay a penalty price for every unit of electricity by which they fall short of their obligation. The mechanism acts to create a market for electricity, allowing competition amongst renewable generators to meet the needs of that market. The underlying theory is that competition in this market place will drive down the costs of supplying renewable electricity and thus minimize the costs to the consumer of meeting renewable energy targets.

The market allows a government to set the capacity that is required, and allows the market place to set the cost. The level at which the penalty price is set allows the government to place an upper limit on the total costs to the consumer.

The mechanism is in place in different forms in just over half of U.S. states as well as the United Kingdom, Italy and Belgium amongst other European countries. In the U.S., quota mechanisms applied at the state level are often assisted by the intermittent application of a federally mandated production tax credit.

Advantages

- Competition: Technologies must compete for contracts, and thus there is pressure to drive down costs and in theory, this means that the costs to the consumer are minimized. There is growing evidence however that this does not always work in practice, with the International Energy Agency suggesting that the less competitive tariffmechanism appears to be able to deliver like for like renewable energy more cheaply than the quota mechanism.
- Static efficiency: It has been suggested that quota mechanisms are more efficient in achieving specific goals for renewable energy capacity.

Disadvantages

- Increased risk: In practice, generators are vulnerable to considerable risks, the effect of which is to raise the cost of capital and thus the overall cost of generation, resulting in an increase in the cost to the consumer. There is evidence that the risk and the related additional costs associated with a quota mechanism are sufficient to outdo any benefits from competition inherent to the system. The specific risks include:
    - Volume (there being no guarantee that they will be able to sell all of their peak electricity, due to competition)
    - Price (as this depends on the market for both electricity and tradeable certificates)
    - Balancing cost (a function of generator intermittency vs. 24-hour base demand)
    - Regulation (i.e. the withdrawal of political support effectively ends the market completely)
- Dynamic efficiency: The mechanism tends to support only those technologies which are close to market when

it is introduced. Technologies outside the mechanism are likely to become less and less competitive, and thus never be developed. Thinking out-of-the-box is effectively penalised. One solution is to provide additional support outside the mechanism, however, this has the disadvantage of raising the overall costs of supporting renewables.

CONTRACT BIDDING MECHANISMS

Governments place an obligation on supply companies to accept electricity from renewable energy generating technologies which have been awarded contracts by government. Generators win these contracts by taking part in a competitive bidding process organized by the government or a specified agent of the government. Historically, competition has usually taken place within technology 'bands', that is, such that competing bids are only compared between generators employing the same technology. For example, wind generator against wind generator. Essentially, this means there are usually different competitions going on at the same time for each technology. Generally, the lowest bids are awarded contracts, provided they meet any criteria set down by the government as part of the competitive process. Examples of such a mechanism in practice include the UK's Non-Fossil Fuel Obligation (NFFO), Ireland's Alternative Energy Requirement (AER) and the French EOLE.

Advantages

- Competition: The mechanism encourages competition, acting to reduce subsidy costs to consumers. Because the governments can limit the number of contracts awarded, they can also control the maximum public spending linked to the scheme.
- Stable pricing: The fixed price attached to successful contracts also mean that project developers have less worries over variations in pricing caused by fluctuations in the electricity market. However, this can also be a problem in fitting the mechanism in with a market-led electricity supply industry.

Disadvantages

- In practice, many contracts awarded under these mechanisms have failed to lead to actual development. There are a number of possible explanations for this. One possibility is that companies make bids against future cost reductions, taking advantage of the preparation period that the mechanisms usually allow before the payment period begins. If the economics of development are not sufficiently attractive towards the end of this period then the project does not go ahead.
- Another possibility is that some companies actually make bids they know will never become economic and will never be developed purely to deny the opportunity for a successful bid to their competitors. The introduction of a penalty for those failing to make good on contracts might act to discourage these problems.

TARIFF MECHANISMS

In a Feed-in Tariff mechanism, the government fixes a price for every unit of electricity produced from any technology which it classifies as renewable. Because fossil fuels are relatively cheap, this price is typically greater than the market price for electricity available in that territory and thus tariffs enable generators to operate economically. Different tariff levels may be set for different technologies. A government may provide the subsidy from its own funds or may compel utility companies to purchase the electricity thus produced, passing the costs on to its consumers. Network supply companies are compelled to accept all electricity from specified technologies.

Advantages

- Risk reduction for investors: Generators receive a fixed price for a fixed period, thus reducing volume and price risk to investors. Generators are also not subject to balancing risk as network companies are compelled to take all electricity. There is also reduced regulatory risk in comparison with other mechanisms in that once a plant is operational it is effectively guaranteed a price for a fixed period into the future.

- Dynamic efficiency: Different technologies develop at different rates. If a government wishes to support a new technology it can provide a tariff specific to that technology and thus encourage it to move closer to market. The balance of evidence suggests that this provides long term benefits in terms of developing more competitive technologies.
- Proven capability: Tariff mechanisms have been widely applied in Europe, and have enjoyed particular success in Germany, Denmark and Spain. Their employment has led to significant increases in renewable electricity generating capacity, particularly of wind energy.

Disadvantages

- Excessive margins for developers: The fixed price over time means that it is difficult to pass on the benefits of increased technological efficiency to consumers. Instead benefits accrue at the level of the generating plant owner, who may be able to access high rates of return. One possible solution is through digression, that is, lowering the tariff rate over time. Reductions in the tariff must be transparent to ensure investor uncertainty is minimized. There is no guarantee that reductions will match the actual improvements in the technology.
- Unpredictable: Whilst tariff mechanisms fix the price available to renewable energy generators, the level of capacity is subject to the market, that is, there is no way of predicting how many investors will be attracted to generation by the price available. This means it is not possible to predict the overall costs of the mechanism in either the short or long term. This can be unattractive to government and consumers/taxpayers.
- Network balancing: Distribution network operators are compelled to accept all electricity from renewable generators, regardless of the demand for electricity at the time of generation. This can lead to network balancing issues, and these tend to increase with the level of intermittent generation on the network. This

leads to increasing potential for technological problems and for increased costs to the network operator.

- Market prioritization: Compelling network operators to accept all renewable generation means that electricity from renewables is always the first to be bought. This effectively interferes with any open market for general electricity generation, and impacts on the ability of "traditional" generators to compete in the electricity sector. This can be problematic where governments are committed to maximizing competition in the markets.

PRODUCTION TAX CREDITS

Production tax credits support the introduction of renewables by allowing companies which invest in renewables to write off this investment against other investments they make. A PTC can be used as the central mechanism for the support of renewables as part of a national or regional mechanism, or it can be used in support of other mechanisms, such as a quota mechanism. Production tax credits have been supplied at the federal level in the U.S.; they have tended to be most effective in states which also provide some other form of support, most notably a quota mechanism.

Advantages

- Stability: Have proven to stimulate capacity alongside quota mechanisms. They may provide a useful way to bring stability to generators when used in this way, reducing uncertainty and thus capital costs.

Disadvantages

- In practice, tax credits tend to be vulnerable to political conflict regarding their maintenance in the long term.
- Tax credits also tend to restrict investment in renewables to large companies with significant portfolios against which they can write off the tax credits they earn. Potential owners of renewable energy projects eligible for tax credits who do not have sufficient tax appetite must seek tax equity investors to partner with in order

to capture this form of support. In the last quarter of 2008, the tax equity market has dried up as a consequence of the global credit crisis, which has resulted in increased cost of energy from renewable energy projects.

- Government "chooses winners and losers" by subsidizing specific technologies, which may prevent the most efficient technology from becoming the most widely adopted.

SCENARIOS AND BLUEPRINTS

Many governments and environmental organisations have published reports outlining a future renewables fuel mix, or position papers on future commitments.

- Greenpeace has published the report "Energy [R]-evolution", details how to halve the world's CO_2 emissions by 2050, using existing technology and still providing affordable energy and economic growth. The report attempts a scenario which would phase out unsustainable energy, and decouple economic growth from fossil fuel use.

Australia

- WWF Australia, together with other members of the Clean Energy Future Group, has published *Clean Energy Future for Australia,* outlines how Australia can meet its electricity needs through a combination of wind, biomass, natural gas and greater energy efficiency. They have also published specific reports for the Australian states of New South Wales, Queensland, Victoria, and Western Australia.

TRENDS FAVOURING RENEWABLES

The renewable market will boom when cost efficiency attains parity with other competing energy sources. The following trends are a few examples by which the renewables market is being helped to attain critical mass so that it becomes competitive enough versus fossil fuels:

Other than market forces, renewable industry often needs government sponsorship to help generate enough momentum in the market. Many countries and states have implemented incentives — like government tax subsidies, partial copayment schemes and various rebates over purchase of renewables — to encourage consumers to shift to renewable energy sources. Government grants fund for research in renewable technology to make the production cheaper and generation more efficient.

Development of loan programs that stimulate renewable favoring market forces with attractive return rates, buffer initial deployment costs and entice consumers to consider and purchase renewable technology. A famous example is the solar loan program sponsored by UNEP helping 100000 people finance solar power systems in India. Success in India's solar program has led to similar projects in other parts of developing world like Tunisia, Morocco, Indonesia and Mexico.

Imposition of high fossil fuel consumption / carbon taxes, and channel the revenue earned towards renewable energy development.

Many think-tanks are warning that the world needs an urgency driven concerted effort to create a competitive renewable energy infrastructure and market. The developed world can make more research investments to find better cost efficient technologies, and manufacturing could be transferred to developing countries in order to use low labor costs. The renewable energy market could increase fast enough to replace and initiate the decline of fossil fuel dominance and the world could then avert the looming climate and peak oil crises.

Most importantly, renewables is gaining credence among private investors as having the potential to grow into the next big industry. Many companies and venture capitalists are investing in photovoltaic development and manufacturing. This trend is particularly visible in Silicon valley, California, Europe, Japan.

ALL ELECTRICITY FROM RENEWABLE SOURCES (AERS)

On July 2008, Al Gore, challenged the United States to commit to producing all electricity from renewable sources

(AERS) like solar and wind power in 10 years. Al Gore´s Alliance for Climate Protection has created the Repower America project to promote this goal. Center for Resource Solutions supports Al Gore's AERS goal.

Scientists from the University of Kassel have been busy proving that Germany can power itself entirely by renewable energy.

Some autonomous regions in Spain lead Europe in the use of renewable energy technology, and plan to reach 100% renewable energy generation in a few years (AERS goal). Castile and León and Galicia are especially near this goal, producing in 2006 70% of their total electricity demand from renewable energy sources, and 5 communities produces more than 50% from renewables.

The Oregon Institute of Technology announced that it was going to be the world's first university to be powered entirely by geothermal energy and the University of Oklahoma has picked up the use-only-renewable-energy gauntlet and has announced that the school's main campus will be entirely powered by wind by 2013 .

Investments

Venture capital and private equity investments in clean energy companies increased by 167 percent in 2006, according to investment analysts at New Energy Finance Limited .

These clean energy investments increased from $2.7 billion in 2005 to $7.1 billion in 2006, driven mainly by a surge of investments in biofuels in the United States. Investments in biofuels more than quadrupled, increasing from $647 million in 2005 to $2.8 billion in 2006.

In addition, investments in solar energy more than tripled, while wind power investments more than doubled. Investments in other clean energy technologies—including energy efficiency, fuel cells, hydrogen, smart power distribution, and carbon markets—grew by 74 percent.

New investment into the sector jumped US$148 billion in 2007, up 60 per cent over 2006, noted a report by the Sustainable Energy Finance Initiative (SEFI). Wind energy attracted one-third of the new capital and solar one-fifth. But interest in solar is growing rapidly on the back of major technological advances which saw solar investment increase 254 per cent . The IEA predicts US$20 trillion will be invested into alternative energy projects over the next 22 years .

STATUTORY DEFINITIONS OF RENEWABLE ENERGY

Statutory definition of renewable energy may differ from a scientific definition, reflecting the legislators' other concerns, such as environmental damage and political sustainability. Contentious cases are, but not limited to: impure biomass or waste materials (such as painted wood), large hydroelectric power plants, and nuclear power. Regarding nuclear energy there is an ongoing public discussion, whether it can or cannot be included in any definition of renewable energy; in any case, there is a demand for a name for the energies that does not includes fossil fuels and nuclear power.

IEA Definition

Though International Energy Agency is not a legislative body, the definition used by this specialised organisation of OECD is representative and influential (... denotes ellipsis from original text):

"Renewables include the following categories:

*Combustible Renewables and Waste** (CRW):

Solid Biomass: *Wood, Wood Waste, Other Solid Waste*: Covers purpose-grown energy crops (poplar, willow etc.), a multitude of woody materials generated by an industrial process (wood/paper industry in particular) or provided directly by forestry and agriculture (firewood, wood chips, bark, sawdust, shavings, chips, black liquor etc.) as well as wastes such as straw, rice husks, nut shells, poultry litter, crushed grape dregs etc.

Charcoal: Covers the solid residue of the destructive distillation and pyrolysis of wood and other vegetal material.

Biogas:: Gases composed principally of methane and carbon dioxide produced by anaerobic digestion of biomass and combusted to produce heat and/or power.

Liquid Biofuels: Bio-based liquid fuel from biomass transformation, ...

Municipal Waste (renewables): Municipal waste energy comprises wastes produced by the residential, commercial and public services sectors and incinerated in specific installations to produce heat and/or power. The renewable energy portion is defined by the energy value of combusted biodegradable material. :Footnote for this item: Some of the waste (the non-biodegradable part of the waste) is not considered renewables as such. However, proper breakdown between renewables and nonrenewables is not always available.

Hydro Power: Potential and kinetic energy of water converted into electricity in hydroelectric plants. It includes large as well as small hydro, regardless of the size of the plants.

Geothermal Energy: Energy available as heat emitted from within the earth's crust, usually in the form of hot water or steam. ...

Solar Energy: Solar radiation exploited for hot water production and electricity generation. Does not account for passive solar energy for the direct heating, cooling and lighting of dwellings or other.

Wind Energy: Kinetic energy of wind exploited for electricity generation in wind turbines.

Tide / Wave / Ocean Energy: Mechanical energy derived from tidal movement, wave motion or ocean current and exploited for electricity generation."

EU Directive 2001/77/EC

Definition of renewable energy sources used by EU legislators coincides with the IEA definition, above. Bio-waste is limited to biodegradable fraction. Article 2. of the EU Directive 2001/77/EC provides the following definition of renewable energy sources:

(2)(a) 'renewable energy sources' shall mean renewable non-fossil energy sources (wind, solar, geothermal, wave, tidal, hydropower, biomass, landfill gas, sewage treatment plant gas and biogases);

further specifying biomass:

(2)(b) 'biomass' shall mean the biodegradable fraction of products, waste and residues from agriculture (including vegetal and animal substances), forestry and related industries, as well as the biodegradable fraction of industrial and municipal waste;

Australia, federal legislation

In Australia, the federal *Renewable Energy (Electricity) Act 2000* takes a broad approach and defines certain energy sources that are eligible renewable energy sources (Act s 17). These include (a) hydro; (b) wave; (c) tide; (d) ocean; e) wind; f) solar; g) geothermal aquifer; h) hot dry rock; (i) energy crops; j) wood waste; k) agricultural waste; l) waste from processing of agricultural products; m) food waste; n) food processing waste; o) bagasse; (p) black liquor; q) biomass based components of municipal solid waste; r) landfill gas; s) sewage gas and biomass based components of sewage; t) any other energy source prescribed by the regulations. Further detail of definitions is found in the Regulations. The Act states s.17(2) that fossil fuels; and materials or waste products derived from fossil fuels are not eligible renewable energy sources. Although the Australian federal legislation does not directly rule out nuclear power as an eligible source, it is not included in the list of eligible sources.

8

Renewable Heat Energy

Renewable heat is an application of renewable energy and it refers to the renewable generation of heat, rather than electrical power (e.g. replacing a fossil fuel boiler using concentrating solar thermal to feed radiators).

Many colder countries consume more energy for heating than electrical power. For example, in 2005 the United Kingdom consumed 354 TWh of electric power, but had a heat requirement of 907 TWh, the majority of which (81%) was met using gas. The residential sector alone consumed a massive 550 TWh of energy for heating, mainly in the form of gas. Almost half of the final energy consumed in the UK (49%) was in the form of heat, of which 70% was used by households and in commercial and public buildings. Households used heat for mainly for space heating (69%) and heating water.

Renewable electric power is becoming cheap and convenient enough to place it, in many cases, within reach of the average consumer. By contrast, the market for renewable heat is mostly inaccessible to domestic consumers due to inconvenience of supply, and high capital costs. Heating accounts for a large proportion of energy consumption, however a universally accessible market is still in its early stages.

LEADING RENEWABLE HEAT TECHNOLOGIES

Solar Heating

Solar heating is a style of building construction which uses the energy of summer or winter sunshine to provide an economic supply of primary or supplementary heat to a structure. The heat can be used for both space heating and water heating (see solar hot water). Solar heating design is divided into two groups:

- Passive solar heating relies on the design and structure of the house to collect heat. Passive solar building design must also consider the storage and distribution of heat, which may be accomplished passively, or use air ducting to draw heat actively to the foundation of the building for storage. One such design was measured lifting the temperature of a house to 24°C (74°F) on a partially sunny winter day (-7°C or 19°F), and it is claimed that the system provides passively for the bulk of the building's heating. The 4,000-square-foot (370 m^2) home cost \$125 per square foot (or 370 m^2 at \$1,351/m^2), similar to the cost of a traditional new home.
- Active solar heating uses pumps to move air or a liquid from the solar collector into the building or storage area. One application, solar water heating, works in conjunction with an existing water heater, and is based on solar panels fitted to the roof. In contrast to photovoltaic panels which are used to generate electricity, solar water heating panels are cheaper to manufacture, and capture a much higher proportion of the sun's energy.

Solar heating systems usually require a small supplementary backup heating system, either conventional or renewable.

Heat Pumps

Heat pumps use work to move heat from one place to another, and can be used for both heating and air conditioning. Though capital intensive, heat pumps are economical to run and can be powered by renewable electricity. Two common types of heat pump are air-source heat pumps (ASHP) and ground-source heat pumps (GSHP), depending on whether heat is transferred from the air or from the ground. Air source heat pumps are not effective when the outside air temperatures is lower than about -15 °C, while ground-source heat pumps are not affected. The efficiency of a heat pump is measured by the coefficient of performance (CoP): For every unit of electricity used to pump the heat, an air source heat pump generates 2.5 to 3 units of heat (i.e. it has a CoP of 2.5 to 3), whereas a GSHP generates 3 to 4 units of heat. Based on current fuel prices for the United Kingdom, assuming a CoP of 3-4, a GSHP can be a cheaper form of space heating than oil, LPG and electric storage heaters. It is however more expensive than mains gas unless the heat pump is linked to an interseasonal thermal store when the CoP can rise to 7 by extracting heat from warm ground .

INTERSEASONAL HEAT TRANSFER

Interseasonal Heat Transfer combines active solar collection to store surplus summer heat in thermal banks with GSHPs to extract it for space heating in winter. This reduces the "Lift" needed and doubles the CoP of the heat pump because the pump starts with warmth from the thermal bank in place of cold from the ground.

CoP and Lift

The CoP increases as the temperature difference, or "Lift", decreases between heat source and destination. The CoP can be maximised at design time by choosing a heating system requiring only a low final water temperature (e.g. underfloor heating), and by choosing a heat source with a high average temperature (e.g. the ground). Domestic Hot Water (DHW) and conventional radiators require high water temperatures, affecting the choice of heat pump technology. Low temperature radiators provide an alternative to conventional radiators.

Wood-pellet Heating

Wood-pellet heating and other types of wood heating systems have achieved their greatest success in heating premises that are off the gas grid, typically being previously heated using heating oil or coal.

Solid wood fuel requires a large amount of dedicated storage space, and the specialised heating systems can be expensive (though grant schemes are available in many European countries to offset this capital cost.) Low fuel costs mean that wood fuelled heating in Europe is frequently able to achieve a payback period of less than 3 to 5 years. Because of the large fuel storage requirement wood fuel can be less attractive in urban residential scenarios, or for premises connected to the gas grid (though rising gas prices and uncertainty of supply mean that wood fuel is becoming more competitive.)

Wood-stove Heating

Burning wood fuel in an open fire is both extremely inefficient (<5%) and polluting due to low temperature partial combustion. In the same way that a drafty building loses heat through loss of warm air through poor sealing, an open fire is responsible for large heat losses by drawing very large volumes of warm air out of the building.

However modern wood stove designs allow for extremely efficient combustion and then heat extraction (>95% efficiency is possible), and draw only small volumes of warm air from the building. High efficiency stoves meet the following design criteria:

- Well sealed and precisely calibrated to draw a low yet sufficient volume of air. Air-flow restriction is critical; a lower inflow of cold air cools the furnace less (a higher temperature is thus achieved). It also allows greater time for extraction of heat from the exhaust gas, and draws less heat from the building.
- The furnace must be well insulated to increase combustion temperature, and thus completeness.
- A well insulated furnace radiates little heat. Thus heat must be extracted instead from the exhaust gas duct. Heat absorption efficiencies are higher when the heat-exchange duct is longer, and when the flow of exhaust gas is slower.
- In many designs, the heat-exchange duct is built of a very large mass of heat-absorbing brick or stone. This design causes the absorbed heat to be emitted over a longer period - typically a day.

Renewable Natural Gas

Renewable natural gas is defined as gas obtained from biomass which is upgraded to a quality similar to natural gas. By upgrading the quality to that of natural gas, it becomes possible to distribute the gas to customers via the existing gas grid. According to the Energy research Centre of the Netherlands, renewable natural gas is 'cheaper than alternatives where biomass is used in a combined heat and power plant or local combustion plant'. Energy unit costs are lowered through 'favourable scale and operating hours', and end-user capital costs eliminated through distribution via the existing gas grid.

Energy Efficiency

Renewable heat goes hand in hand with energy efficiency. Indeed renewable heating projects depend heavily for their success on energy efficiency; in the case of solar heating to cut reliance on the requirement supplementary heating, in the case of wood fuel heating to cut the cost of wood purchased and volume stored, and in the case of heat pumps to reduce the size and investment in heat pump, heat sink and electricity costs.

Two main types of improvement can be made to a building's energy efficiency:

Insulation

Improvements to insulation can cut energy consumption greatly, making a space cheaper to heat and to cool. However existing housing can often be difficult or expensive to improve. Newer buildings can benefit from many of the techniques of superinsulation. Older buildings can benefit from several kinds of improvement:

- Solid wall insulation: A building with solid walls can benefit from internal or external insulation. External wall insulation involves adding decorative weather-proof insulating panels or other treatment to the outside of the wall. Alternatively, internal wall insulation can be applied using ready-made insulation/plaster board laminates, or other methods. Thicknesses of internal or external insulation typically range between 50 and 100 mm.

- Cavity wall insulation: A building with cavity walls can benefit from insulation pumped into the cavity. This form of insulation is very cost effective.
- Programmable thermostats allow heating and cooling of a room to be switched off depending the time, day of the week, and temperature. A bedroom, for example, does not need to be heated during the day, but a living room does not need to be heated during the night.
- Roof insulation
- Insulated windows and doors
- Draught proofing

Underfloor Heating

Underfloor heating is significantly more energy efficient that traditional methods of heating:

- Water circulates within the system at low temperature (35 °C - 50 °C) making gas boilers, wood fired boilers, and heat pumps significantly more efficient.
- Rooms with underfloor heating are cooler near the ceiling, where heat is not required, but warmer underfoot, where comfort is most required.
- Traditional radiators are frequently positioned underneath poorly insulated windows, heating them unnecessarily.

SUSTAINABLE ENERGY

Sustainable energy is the provision of energy such that it meets the needs of the present without compromising the ability of future generations to meet their needs. A broader interpretation may allow inclusion of fossil fuels and nuclear fission as transitional sources while technology develops, as long as new sources are developed for future generations to use. A narrower interpretation includes only energy sources which are not expected to be depleted in a time frame relevant to the human race.

Sustainable energy sources are most often regarded as including all renewable sources, such as biofuels, solar power, wind power, wave power, geothermal power and tidal power. It usually also includes technologies that improve energy efficiency. Conventional fission power is sometimes referred to as sustainable, but this is controversial politically due to concerns about peak uranium, waste disposal and the risks of disaster due to accident, terrorism, or natural disaster.

DISTINCTION FROM OTHER TERMS

Some ways in which *sustainable energy* has been defined are:

- "Effectively, the provision of energy such that it meets the needs of the future without compromising the ability of future generations to meet their own needs. ...Sustainable Energy has two key components: renewable energy and energy efficiency." – *Renewable Energy and Efficiency Partnership* (British)
- "Energy which is replenishable within a human lifetime and causes no long-term damage to the environment." – *Jamaica Sustainable Development Network*

This sets *sustainable energy* apart from other renewable energy terminology such as *alternative energy* and *green energy*, by focusing on the ability of an energy source to continue providing energy. Sustainable energy can produce some pollution of the environment, as long as it is not sufficient to prohibit heavy use of the source for an indefinite amount of time. Sustainable energy is also distinct from Low-carbon energy, which is only sustainable as in not adding lots of CO_2 to the atmosphere.

RENEWABLE ENERGY TECHNOLOGIES

Renewable energy technologies are essential contributors to sustainable energy as they generally contribute to world energy security, reducing dependence on fossil fuel resources, and providing opportunities for mitigating greenhouse gases. The International Energy Agency has defined three generations of renewable energy technologies, reaching back more than 100 years:

- *First-generation technologies* emerged from the industrial revolution at the end of the 19th century and include hydropower, biomass combustion, and geothermal power and heat. Some of these technologies are still in widespread use.

- *Second-generation technologies* include solar heating and cooling, wind power, modern forms of bioenergy, and solar photovoltaics. These are now entering markets as a result of research, development and demonstration (RD&D) investments since the 1980s. The initial investment was prompted by energy security concerns linked to the oil crises of the 1970s but the continuing appeal of these renewables is due, at least in part, to environmental benefits. Many of the technologies reflect significant advancements in materials.
- *Third-generation technologies* are still under development and include advanced biomass gasification, biorefinery technologies, concentrating solar thermal power, hot dry rock geothermal energy, and ocean energy. Advances in nanotechnology may also play a major role.

First- and second-generation technologies have entered the markets, and third-generation technologies heavily depend on long term research and development commitments, where the public sector has a role to play.

A 2008 comprehensive cost-benefit analysis review of energy solutions in the context of global warming and other issues ranked wind power combined with battery electric vehicles (BEV) as the most efficient, followed by concentrated solar power, geothermal power, tidal power, photovoltaic, wave power, coal capture and storage, nuclear energy, and finally biofuels.

FIRST-GENERATION TECHNOLOGIES

First-generation technologies are most competitive in locations with abundant resources. Their future use depends on the exploration of the available resource potential, particularly in developing countries, and on overcoming challenges related to the environment and social acceptance. Among sources of renewable energy, hydroelectric plants have the advantages of being long-lived — many existing plants have operated for more than 100 years. Also, hydroelectric plants are clean and have few emissions. Criticisms directed at large-scale hydroelectric plants include: dislocation of people living where the reservoirs are planned, and release of significant amounts of carbon dioxide during construction and flooding of the reservoir. However, it has been found that high emissions are associated only with shallow reservoirs in warm (tropical) locales. Generally speaking, hydroelectric plants produce much lower life-cycle emissions than other types of generation. Hydroelectric power, which underwent extensive development during growth of electrification in the 19th and 20th centuries, is experiencing resurgence of development in the 21st century. The areas of greatest hydroelectric growth are the booming economies of Asia. China is the development leader; however, other Asian nations are installing hydropower at a rapid pace. This growth is driven by much increased energy costs — especially for imported energy — and widespread desires for more domestically-produced, clean, renewable, and economical generation.

Geothermal power plants can operate 24 hours per day, providing base-load capacity, and the world potential capacity for geothermal power generation is estimated at 85 GW over the next 30 years. However, geothermal power is accessible only in limited areas of the world, including the United States, Central America, Indonesia, East Africa and the Philippines. The costs of geothermal energy have dropped substantially from the systems built in the 1970s. Geothermal heat generation can be competitive in many countries producing geothermal power, or in other regions where the resource is of a lower temperature.

SECOND-GENERATION TECHNOLOGIES

Markets for second-generation technologies are strong and growing, mainly in countries such as Germany, Spain, the United States, and Japan. The challenge is to broaden the market base for continued growth worldwide. Strategic deployment in one country not only reduces technology costs for users there, but also for those in other countries, contributing to overall cost reductions and performance improvement.

Solar heating systems are a well known second-generation technology and generally consist of solar thermal collectors, a fluid system to move the heat from the collector to its point of usage, and a reservoir or tank for heat storage and subsequent use. The systems may be used to heat domestic hot water, swimming pool water, or for space heating. The heat can also be used for industrial applications or as an energy input for other uses such as cooling equipment. In many climates, a solar heating system can provide a very high percentage (50 to 75%) of domestic hot water energy.

In the 1980s and early 1990s, most photovoltaic modules provided Remote Area Power Supply, but from around 1995, industry efforts have focused increasingly on developing building integrated photovoltaics and power plants for grid connected applications (see photovoltaic power stations article for details). Currently the largest photovoltaic power plant in North America is the Nellis Solar Power Plant (15 MW). There is a proposal to build a Solar power station in Victoria, Australia, which would be the world's largest PV power station, at 154 MW. Other large photovoltaic power stations, which have been proposed or are under construction, include the Girassol solar power plant (62 MW), and the Waldpolenz Solar Park (40 MW).

Some of the second-generation renewables, such as wind power, have high potential and have already realised relatively low production costs. At the end of 2006, worldwide capacity of wind-powered generators was 74,223 megawatts, and although it currently produces less than 1% of world-wide electricity use, it accounts for approximately 20% of electricity use in Denmark, 9% in Spain, and 7% in Germany. However, it may be difficult to site wind turbines in some areas for aesthetic or environmental reasons, and it may be difficult to integrate wind power into electricity grids in some cases.

Solar thermal power stations have been successfully operating in California commercially since the late 1980s, including the largest solar power plant of any kind, the 350 MW Solar Energy Generating Systems. Nevada Solar One is another 64MW plant which has recently opened. Other parabolic trough power plants being proposed are two 50MW plants in Spain, and a 100MW plant in Israel. Brazil has one of the largest renewable energy programs in the world, involving production of ethanol fuel from sugar cane, and ethanol now provides 18 percent of the country's automotive fuel. As a result of this, together with the exploitation of domestic deep water oil sources, Brazil, which years ago had to import a large share of the petroleum needed for domestic consumption, recently reached complete self-sufficiency in oil.

Most cars on the road today in the U.S. can run on blends of up to 10% ethanol, and motor vehicle manufacturers already produce vehicles designed to run on much higher ethanol blends. Ford, DaimlerChrysler, and GM are among the automobile companies that sell "flexible-fuel" cars, trucks, and minivans that can use gasoline and ethanol blends ranging from pure gasoline up to 85% ethanol (E85). By mid-2006, there were approximately six million E85-compatible vehicles on U.S. roads.

THIRD-GENERATION TECHNOLOGIES

Third-generation technologies are still under development and include advanced biomass gasification, biorefinery technologies, solar thermal power stations, hot dry rock geothermal energy, and ocean energy. Third-generation technologies are not yet widely demonstrated or have limited commercialization. Many are on the horizon and may have potential comparable to other renewable energy technologies, but still depend on attracting sufficient attention and RD&D funding. According to the International Energy Agency, new bioenergy (biofuel) technologies being developed today, notably cellulosic ethanol biorefineries, could allow biofuels to play a much bigger role in the future than previously thought. Cellulosic ethanol can be made from plant matter composed primarily of inedible cellulose fibers that form the stems and branches of most plants. Crop residues (such as corn stalks, wheat straw and rice straw), wood waste, and municipal solid waste are potential sources of cellulosic biomass. Dedicated energy crops, such as switchgrass, are also promising cellulose sources that can be sustainably produced in many regions of the United States. In terms of Ocean energy, another third-generation technology, Portugal has the world's first commercial wave farm, the *Agucadora Wave Park*, under construction in 2007. The farm will initially use three Pelmis P-750 machines generating 2.25 MW. and costs are put at 8.5 million euro. Subject to successful operation, a further 70 million euro is likely to be invested before 2009 on a further 28 machines to generate 525 MW. Funding for a wave farm in Scotland was announced in February, 2007 by the Scottish Executive, at a cost of over 4 million pounds, as part of a £13 million funding packages for ocean power in Scotland. The farm will be the world's largest with a capacity of 3 MW generated by four Pelamis machines. (see also Wave farm).

In 2007, the world's first turbine to create commercial amounts of energy using tidal power was installed in the narrows of Strangford Lough in Ireland. The 1.2 MW underwater tidal electricity generator takes advantage of the fast tidal flow in the lough which can be up to 4m/s. Although the generator is powerful enough to power up to a thousand homes, the turbine has a minimal environmental impact, as it is almost entirely submerged, and the rotors turn slowly enough that they pose no danger to wildlife.

Solar power panels that use nanotechnology, which can create circuits out of individual silicon molecules, may cost half as much as traditional photovoltaic cells, according to executives and investors involved in developing the products. Nanosolar has secured more than $100 million from investors to build a factory for nanotechnology thin-film solar panels. The company's plant has a planned production capacity of 430 megawatts peak power of solar cells per year. Commercial production started and first panels have been shipped to customers in late 2007.

Energy Efficiency

Moving towards energy sustainability will require changes not only in the way energy is supplied, but in the way it is used, and reducing the amount of energy required to deliver various goods or services is essential. Opportunities for improvement on the demand side of the energy equation are as rich and diverse as those on the supply side, and often offer significant economic benefits.

Renewable energy and energy efficiency are sometimes said to be the "twin pillars" of sustainable energy policy. Both resources must be developed in order to stabilize and reduce carbon dioxide emissions. Efficiency slows down energy demand growth so that rising clean energy supplies can make deep cuts in fossil fuel use. If energy use grows too fast, renewable energy development will chase a receding target. Likewise, unless clean energy supplies come online rapidly, slowing demand growth will only begin to reduce total emissions; reducing the carbon content of energy sources is also needed. Any serious vision of a sustainable energy economy thus requires commitments to both renewables and efficiency.

Renewable energy (and energy efficiency) are no longer niche sectors that are promoted only by governments and environmentalists. The increased levels of investment and the fact that much of the capital is coming from more conventional financial actors suggest that sustainable energy options are now becoming mainstream. Climate change concerns coupled with high oil prices and increasing government support are driving increasing rates of investment in the sustainable energy industries, according to a trend analysis from the United Nations Environment Programme. According to UNEP, global investment in sustainable energy in 2007 was higher than previous levels, with $148 billion of new money raised in 2007, an increase of 60% over 2006. Total financial transactions in sustainable energy, including acquisition activity, was $204 billion. Investment flows in 2007 broadened and diversified, making the overall picture one of greater breadth and depth of sustainable energy use. The mainstream capital markets are "now fully receptive to sustainable energy companies, supported by a surge in funds destined for clean energy investment".

Nuclear Power

It is said that nuclear has the potential to be sustainable, however, this is often qualified with the argument that there are serious challenges that must be dealt with before it can drastically increase its role. There are two types of nuclear power. "Fission" is used in all current nuclear power plants. Fusion power is the reaction that powers stars, including the sun, which is still being researched for use on earth. Both types create radioactive waste in the form of activated structural material, which is one of the sustainability issues.

Fission power's long-term sustainability depends on the amount of uranium and thorium that are available to be mined, on the operators' abilities safely to dispose of the waste and on the continued prevention of major accidents. Estimates for fuel reserves vary widely, but if breeder reactors and fuel reprocessing are assumed, estimates tend to be tens of thousands of years or longer (uranium is approximately as common in Earth's crust as tin or zinc (2 ppm), and thorium as common as lead (6 ppm)).

Fusion power's long-term sustainability depends on whether or not affordable technology can be developed, and on the amount of lithium that is available to be mined (for deuterium-tritium fusion), or the amount of deuterium, a hydrogen isotope, available in "heavy water" (for deuterium-deuterium fusion). Lithium is a reasonably common component of Earth's crust, being about 10 times as common as thorium (65 ppm). Deuterium occurs in small concentrations wherever hydrogen is found (principally in water), at about 150 ppm. It can be extracted easily from seawater. Given the size of earth's oceans, economically viable reserves of deuterium are for practical purposes unlimited.

TECHNICAL SUSTAINABILITY OF NUCLEAR POWER

Proponents, such as former U.S. Environmental Protection Agency administrator Christine Todd Whitman and Greenpeace co-founder Patrick Moore (both co-chairs of the Clean and Safe Energy Coalition) also claim that nuclear power is at least as environmentally friendly as traditional sources of renewable energy, making it part of the solution to global warming and the world's growing need for energy. They note that nuclear power plants, once built and before decommissioning begins, produce little carbon dioxide emissions and point out that the radioactive waste produced is minimal and well-contained, especially compared to fossil fuels.

9

Deforestation and Forest Degradation by Commercial Harvesting for Firewood and Charcoal

INTRODUCTION

In Nicaragua, the Pacific region is home to a forest ecosystem adapted to low rainfall (800-1200 mm/year) distributed over a period of six months (May to November) which is followed by a six months dry period (December to April). As a consequence of the dry season, most trees are deciduous, losing their leaves during this time. The appearance of this forest during the dry season is similar to that of a temperate forest during winter, trees without leaves, in contrast to the evergreen tropical rain forest.

Since colonial times, the tropical dry forest of Nicaragua has been gradually diminishing (Table 1). In the last fifty years, with the development of commercial agriculture (cotton, sugar cane and coffee), a major change in the use of land has occurred; the needs for firewood and construction (roads and cities) has further increased the pressure over the tropical dry forest.

IMPACT FROM FIREWOOD AND CHARCOAL PRODUCTION

In 1998 PROLEPA, an NGO whose aims are to promote the development and modernization of bioenergy in Nicaragua and Honduras, carried out a preliminary survey in five traditional fuelwood production sites in the Pacific region, to highlight the negative impacts on the environment, particularly forest degradation and deforestation. It did this study with the support of ESMAP (the Energy Sector Management Assistance Program), a joint World Bank-UNDP programme.

Eight square kilometres of land, located in five sites in the region, were analyzed for both deforestation and forest degradation from 1986 to 1996. Particular importance was given to observe the linkages between changes of land use and fuelwood harvesting. It was found that deforestation appears to be more associated with the changes of use of land from forest to agriculture and cattle ranching whilst forest degradation appears to be directly associated with fuelwood harvesting for both household and industrial firewood and charcoal production.

Table 1: Difference between annual rate of wood harvesting and natural wood regeneration for two types of forest in San Pablo community, Nicaragua

Type of forest	Harvesting cycle (years)	Amount of wood harvested (m3/ha)	Rate of harvesting (m3/ha/yr)	Reduction per year (m3/ha/yr)
Savannah	3	15.08	5.02	1.5
Secondary dry forest	3	30.16	10.05	2.5

Also, in a more recent unpublished survey implemented by PROLEÑA within EMOLEP (Sustainable Fuelwood Supply and Efficient Demand Project Strategy for the Pacific region), there were again indications that firewood is not necessarily the main cause for either deforestation or forest degradation in the Pacific region. In areas in which primary or very dense tropical dry forest are being harvested for commercial firewood according to a management plan, the harvesting levels were found to be below the forest production capacity, e.g. not jeopardizing the sustainability of the forest ecosystems. However, these primary or very dense tropical dry forests are today the exception rather than the rule. There are indications that if this very dense forest has already being harvested and degraded non-sustainably by timber harvesters, continuously harvesting for firewood contributes a great deal to the further degradation and eventual deforestation of the site.

Forest Regulations

Although there are forest regulations in Nicaragua which define the legal and technical procedures needed to obtain a permit to harvest fuelwood, enforcement is not a general rule. These regulations, if well understood and applied in the field, provide an excellent tool to guide landowners towards the sustainable management of forest resources, guaranteeing a maximum productivity over a longer period of time, and with little or no negative impact over the forest ecosystem.

In the case of the Pacific region of Nicaragua, lack of planning has contributed to a degradation of the forest resources resulting in a significant reduction in forest productivity, loss of biodiversity (including many tree species, other plants and wildlife), loss or reduced watershed capability, and soil erosion.

CAUSES OF DEFORESTATION AND FOREST DEGRADATION

The causes for a non-sustainable management of the dry forest are many, but the main ones are:

- Lack of technical knowledge in how to manage the dry forest: unfortunately, at present, the appropriate knowledge to manage the dry forest effectively is not well known among landowners and most of the foresters.
- Lack of long term security over the forest resource: The National Constitution defines that all forest resources belong to the nation, creating uncertainty over the government rights to confiscate trees in private lands. Further, too many landowners do not have proper land titles (either due to the high cost of the paperwork, or because of land property rights disputes), creating uncertainty over the legal rights for future land ownership.
- Lack of information about forest products: Landowners usually do not have adequate knowledge about the broad range of products and services that his or her forest can

provide to society, they also have insufficient information about markets and prices.

- Lack of awareness: Environmental awareness is not yet well established among landowners in Nicaragua, although there is a small but growing awareness.
- Lack of economical incentives: There are no economic incentives for landowners to manage or protect the natural dry forests properly. Fuelwood prices are very low, around US$ 6/tonne, which does not provide any incentive for landowners to reinvest in management of the forest; fuelwood prices do not reflect the cost of replanting. There are no government-funded incentive schemes for natural forest management, such as tax exemptions, and no resources are committed to forest protection or even reforestation.
- Lack of adequate control: The governmental regulating agencies such as the National Forest Institute and municipal governments have insufficient resources, such as personnel and transport, to control and monitor the use of the natural forests effectively.
- Lack of a positive government attitude: In general, the government agencies do not have a positive attitude toward the fuelwood sector, seeing them as depredators and difficult to control. The fuelwood sector comprises people with little environmental and technical awareness, who receive the lowest pay in the forestry sector. Government policies and attitudes restrict the sector with many regulations and excessive taxes. They do not understand the need to provide the fuelwood sector with economic incentives, less bureaucracy and general assistance towards increased profitability.

Hopes for Sustainability

In spite of the above, there are signs that the deforestation and degradation of the tropical dry forest of Nicaragua by the fuel-wood sector is about to be stopped. In the most recent surveys in the region, there are a few examples of landowners, with financial support from fuel-wood merchants, that are implementing successful sustainable forest management plans. This indicates that it is possible to achieve rational use of the forest and that a partnership with those most interested in sustaining the fuelwood business, such as fuel merchants, is possible. The challenge will be disseminating these examples so that they become the norm. Also, a few of the largest fuelwood consuming industries are now establishing or seriously considering how to establish fast-growing fuelwood plantations, and this will definitely reduce pressure over the natural forest.

PROLEÑA is helping to modernize the fuelwood sector of Nicaragua. It is attracting the attention of international donors through its work, including The World Bank, The Inter-American Development Bank (IDB) and The Food and Agriculture Organization of the United Nations (FAO). A proposed course of action being negotiated with such donors, includes a mechanism for providing technical and financial incentives to landowners to achieve sustainable forest management, establishing multiple purpose plantations, developing a new less bureaucratic and effective legislation, and reducing household fuelwood demand by promoting improved woodstoves and fuelwood substitution. This last action may prove very important for Nicaragua, since the latest PROLEÑA household survey indicates that over 50% of the Nicaraguan population uses fuel-wood for cooking, and within this group the majority uses traditional woodstoves such as open fires (three stones) or semi-closed (U-shaped) stoves.

With technical support from PROLEÑA, the National Energy Commission, a governmental executive agency, is developing the EMOLEP project. The strategic plan will include supporting management of the natural forests, fuelwood plantations, improved woodstoves and even fuelwood substitution. This plan will be implemented by coordinating public and private organizations, and will be presented to the international donors for financial and technical support.

The information in this article indicates that Nicaragua is perhaps at a turning point in its history of the use of fuelwood: from a history of unsustainable and wasteful use of the fuelwood, toward a future of more sustainable and efficient use of the fuel-wood.

CONCLUSIONS

Fuelwood harvesting for firewood and charcoal is not the main factor causing the deforestation in the tropical dry forest of the Pacific region of Nicaragua. The usual causes of deforestation are clearing of land for agriculture and pastures. Timber harvesting, cattle ranching or even fuelwood collection in forests already degraded by other forms of fuelwood production can contribute to further forest degradation.

Fuelwood harvested in primary or dense forest under a management plan can be done commercially within sustainable levels. Reforestation with fast-growing multiple-purpose trees by larger fuelwood consuming industries can reduce, by a significant level, the pressure on the natural dry forest of Nicaragua.

The laws on forestry and the government agencies in Nicaragua will neither control the fuelwood sector effectively nor provide incentives to achieve sustainability. However, there are signs that management of the natural forest for fuelwood is possible and that the fuelwood sector is concerned about its future and is taking initiatives to modernize it.

EFFECT OF EXPANDING SUGAR-CANE FARMING ON COMMUNITY WOODFUEL

In most developing countries, dependence on biomass fuel as a source of energy is very high, especially in rural areas; Uganda is no exception. Over 87% of the population in Uganda live in rural areas (1) and are reliant for nearly all their energy needs on biomass (2). The availability of biomass fuel depends largely on how land is being utilized. Land use changes are therefore a crucial factor in determining the availability of fuelwood at household level. Commercial agriculture expansion has a large impact on the locations from which local people collect their woodfuel.

Masindi District

Over the last couple of years dramatic changes in land use have occurred in rural areas in Uganda. In 1972, the dictator Idi Amin forced the Asians to leave Uganda at short notice, leaving behind all their possessions. Particularly in rural areas, large estates were left fallow and, to some extent, were occupied by the local population. The current Museveni regime welcomes ousted Asians to reclaim their land.

Two sub-counties of the Masindi district were regions of severe fighting in the 1970s. Besides the land left behind by ousted Asians, the state-owned Kinyara sugar estate was abandoned. Large areas were left unattended and were also a major source of woodfuel for the local population.

With the return of peace in Uganda, a start was made to reactivate the Kinyara sugar estate. At present, the estate is managed by an international consortium, Kinyara Sugar Works, which reclaimed the land and planted sugar-cane. Besides the Kinyara-owned land (the so-called nucleus estate), the factory has now started to work with out-growers.

Outgrowers are local farmers growing sugar-cane on their land aided by the estate, which clears bushes and trees from the land, ploughs it, plants young cane and harvests the yield. Farmers assisted in this way live within a 10 km radius of the nucleus sugar estate. Recently the area in which Kinyara has facilitated outgrowers was extended to a 15 km radius. An increase to 20 km can be expected in the near future (3).

RESPONSES OF THE LOCAL POPULATION

The area under study lies between the protected Budongo Forest Reserve and the current sugar estate. Villages in this area can be grouped into three different groups:

1. the area next to the sugar-cane fields
2. the area adjacent to the protected forest
3. the area in between areas 1 and 2.

From each region, one village was selected in which household interviews were conducted to investigate the current and past situation regarding the source of biomass energy used. The villages of Kabango, Nyabyeya and Nyantonzi were selected respectively. Table 1 gives an overview of the results of the household interview conducted in July 1998.

Table 1 Percentage of total number of respondents collecting woodfuel from certain land uses now and in the past

Region	Village	Year	Natural woodlands	Private farms	Private fallow land	Protected riverine forest	Sugar-cane cleared area	Average distance
1	Kabango	1998	12	46	19	19	12	1.5
		1960	46	31	23	0	-	1.0
2	Nyabyeya	1998	4	0	15	89	0	1.1
		1960	41	4	0	30	-	0.6
3	Nyantonzi	1998	48	48	0	44	0	1.9
		1960	56	36	16	4	-	1.6

The village in the first region (Kabango) showed very clear evidence of wood scarcity. As shown in Table 1, villagers used to collect their fuel wood mainly from natural woodlands, which are now converted into sugar-cane growing areas. At present, nearly 50% of the households in Region 1 buy charcoal from traders who use wood from land being cleared for sugar-cane growing. This source is temporary and will cease to exist when all the potential sugar-cane growing area has been cleared.

The village near to the protected forest (Nyabyeya) shows a shift in collecting fuel from natural woodlands to protected forest resources. Analysis indicates that the original fuel wood collecting area of woodland is now in used for subsistence farming, leaving the protected forest as the only nearby source. Access to this resource is still permitted, but government policies might alter this situation.

The third village (Nyantonzi), in the intermediate area, also shows a much higher dependence on the protected forest reserve but, unexpectedly, no reliance on the cleared sugar-cane area.

LAND USE CHANGE ANALYSIS

Land cover maps were produced for the situation in I960 and in 1998. The I960 land use map (Figure 1) was based on an existing map of I960. Classification of land use in the table was based on any information provided on the map. Although incomplete, it clearly shows the sugar-cane growing area as well as forested areas.

The 1998 land-use map was based on the available land use map of 1996, as produced by the National Biomass Study, updated by mapping new sugar-cane growing areas with a handheld GPS satellite receiver. Newly cleared wooded areas were mapped in the same way. It can be seen that the land lost in the woodland and grassland categories nearly equal the increase in commercial and subsistence farming.

An intermediate map for the year 1987 was also produced by estimates based on the answers received from the interviews. The purpose of this map is to determine the expansion of sugar-cane growing fields. The three maps together clearly show that the major land use changes took place in the last decade only.

Concluding Remarks

The data collected in this study show clear evidence of a rapid expansion of the sugar-cane growing area. As a result, local people near the sugar-cane estates have had to go to different areas from those they were using previously to collect their wood. In the current situation, local woodfuel demand can be met by using wood from land cleared in preparation for sugar-cane growing. This supply is not sustain-able. In order to facilitate local wood energy planning, more research will be needed to quantify the demand of wood by the local people and the available biomass in the area.

(a) as percentage of original area of land cover class 1

Possible interventions may be triggered by increasing scarcity of woodfuels and by the re-activation of improved stoves programmes in the area. The Nyabyeya Forestry College in Masindi might play a pivotal role in this by integrating the promotion and utilization of their pekope stove as part of their educational programme.

To solve future fuelwood problems and to conserve the protected Budongo forest it is recommended that fuel-wood plantations are established as a buffer between the forest and villages. Indigenous fast growing species which can be coppiced, as already planted by farmers around their homesteads, might be planted in participatory forest management projects on the degraded areas around Budongo forest.

A rapidly accessible new source of wood fuel for local people could be formed by the small patches of eucalyptus that are planted by Kinyara on their estate in areas not suitable for sugar-cane. Originally these trees were planted to discourage illegal homes being built on the land. Using the trees as woodfuel will give them a more positive function.

Table: Area of land cover changes from year 1960 to 1998. (Area figures x 1000 ha)

Land Cover	Land cover type	Area	Area	Area and nature of change	% change
1	Fully stocked dense tropical high forest	27.6	26.4	Decreased by 1.2	-4.3
2	Degraded tropical high forest/woodland	0.0	0.7	Tropical high forest degraded by 0.7	2.5(a)
3	Plantation (mixed Forest)	0.0	0.3	New cover type 0.3	-
4	Woodland	21.0	14.7	Decreased by 6.3	-30.0
5	Grassland	10.0	6.2	Decreased by 3.8	-38.0
6	Commercial Farmland (sugarcane)	0.7	6.2	Increased by 5.5	78.0
7	Subsistence Farmland	14.0	18.4	Increased by 4.4	31.4
8	Wetlands	0.4	0.4	Remained the same	0.0
9	Built up area	0.1	0.3	Increased by 0.2	200.0
	Total area	73.8	73.8		

Furthermore, government policies have to be formulated that compel companies involved in wood depletion to contribute directly to the replenishment of wood resources.

This article is based on Mugisha, C.H. (1999) Impact of land use change on fuel wood collecting areas: application of remote sensing and GIS. A case study for Budongo and Biiso sub-counties Masindi district Uganda. MSc thesis ITC Forest Science Division, Enschede, the Netherlands.

WORKSHOP REPORT ON URBAN WASTE AND ENERGY IN DEVELOPING COUNTRIES

This workshop was organized by Intermediate Technology' in the UK as part of a project 'Energy' Provision to the Urban Poor', funded by the Department for International Development (DFID). Further information can be obtained from Simon Dunnet or Alison Doig, ITDG, Schumacher Centre for Technology and Development, Bourton Hall, Bourton-on-Dunsmore, Warwickshire CV23 9QZ, UK

Energy supply and demand in urban areas of developing countries is not understood as thoroughly as energy systems in rural areas, despite the fact that the rural economy in areas close to towns and especially in periurban areas, is often dependent on the sale of wood and charcoal.

In the case of large cities, the economy up to 100 km away or more may be affected. The poorest in urban areas continue using traditional fuels because their access to modern energy, such as kerosene, LPG and electricity is limited. Price and supply fluctuations of these modern fuels also mean that households will switch back to using biomass-derived fuels when it is more convenient or cheaper to do so.

The generation of waste in towns and cities could provide a source of fuel to which the urban poor have access. At present, large numbers are employed in both the so-called 'formal' and 'informal' sectors, recycling useful materials that have been discarded by others. The purpose of the workshop was to investigate the possibility of using some of this material as a fuel, and/or those elements that would not otherwise be recycled.

PRESENTATIONS

Informal Sector Recycling

Dr Ali described the contribution that the informal sector makes in the recycling of waste, specifically in Karachi. It is estimated that 50000 people earn their livelihood from this activity, two fifths of them engaged in waste picking. Some key issues were raised during the presentation:

- Why do people separate waste?
- What is being separated?
- What improvements could be made?
- What role should institutions like municipal corporations take?
- What is the role for NGOs?
- How could micro-enterprises be developed around 'Waste-to-energy'?
- Is there a need to look at the whole concept under the social, institutional and financial framework or do we need the most appropriate technology?
- Who will do the operation and maintenance?
- The issue of sustainability.

EXPERIENCE OF URBAN WASTE AND ENERGY IN SRI LANKA

IT Sri Lanka's recent experience in utilizing waste from urban and periurban areas for the production of energy was outlined in the next presentation. It is estimated that there is the potential to generate 101 MW using urban waste in Sri Lanka.

The Sri Lanka office has been involved in the energy sector, and particularly biogas, for twenty years. There are three main biogas projects:

The Kirullapone pilot project uses waste provided by the municipal council to fuel five 60 cubic meter digesters.

The Maharangama pilot project uses waste that has been sorted at the household level to remove most inorganic material (or items that may have some value).

The Anuradhapura pilot project supplies gas to a school hostel for cooking, using partially sorted waste.

Energy for Urban Poor in Developing Countries-Potential Solutions from Waste Materials

The nature of the energy flows within Bangalore, the fastest growing metropolis in India, with reference to the potential for the use of waste, was outlined by Professor Reddy. In this city, the poor use biomass fuels such as fuelwood, charcoal and agriculture wastes (with efficiencies as low as 10%) and the rich use Liquefied Petroleum gas (LPG) and electricity (with efficiencies greater than 60%). The percentage income a household spends on its energy needs reveals that low-income households who use biomass spend as much as 17% while those with high income only spend 0.4%.

Studies on providing energy in Bangalore shows that the municipal waste could not only produce energy but also nitrogen-rich manure. It is estimated that there is 690 tons per day of garbage available in Bangalore of which 50% is manure. Using the fermentable matter there is a potential for 512 MWh of thermal energy or 114 MWh of electrical energy and 340 tons of fertiliser.

The benefits of the adoption of waste energy technologies would be conservation of resources and protection of the global environment. To government, capital investment would be reduced with a corresponding reduction in foreign exchange outflows. The customer gains by being able to satisfy basic needs, reduce costs, and have an improved quality of service.

Building Materials and Energy: Possibilities and Alternatives for Developing Countries

The use of ash from waste materials combustion for the production of lime-pozzolana binders was the initial reason that Dr Martirena was interested in the energy aspects of building material production. The main sources of waste for this purpose are agro-industrial waste and solid municipal waste. Biomass is processed and densified with simple hardware into a solid fuel block. Heat from its combustion can be used for drying, for the production of steam and for quicklime burning. Flue gases can be filtered to produce gypsum, which, along with the ashes, can be used to manufacture the lime-pozzolana binders. The main product is a lime-pozzolana binder produced at a very small scale. The technology has been under development for the last 8 years, and it is just at the stage of transferring from pilot scale applications into commercial applications (The EU has recently approved a project to spread lime-pozzolana workshops around Central America). There are a few of these workshops in commercial operation.

Key Issues for Sorting Waste for Energy

Following the presentations, the participants were asked to discuss the question 'what are the key issues for waste management in the informal sector of developing countries, with emphasis on sorting wastes for energy recovery? Key areas were identified as:

- Research into how linkages between the formal/informal, private/public sectors could be established,
- Collaboration should be encouraged between workshop participants,
- Lesson-learning and south/south exchanges should be encouraged,
- Urban waste-to-energy technologies need investigation,
- Suggestions to benefit the poor should be made to policy makers
- Knowledge transfer within the informal sector regarding technologies needs further understanding.

One overriding theme of the workshop was that the production of energy from urban waste should form part of a wider poverty alleviation process, rather than the development of technology for its own sake. If the shortage of, or limited access to, sources of energy for the urban poor are taken as the starting point, then the development or adaptation of structures and technologies to utilize waste could occur in a sustainable way.

Concern was expressed about the fragile nature of this informal livelihood from waste picking, in particular that any intervention should benefit these groups, and not remove their access to the waste. There is a great need to understand and to support the informal waste-recycling sector. Environmental health was a key issue raised throughout the workshop. Hazards from uncollected, often putrefied wastes were highlighted. The health risks to the people involved with handling and sorting of waste was also raised.

The workshop identified both a need and opportunities for further activities in this field. The first steps will be to:

- increase understanding of the informal waste sorting systems,
- address the environmental health risks of informal waste management,
- undertake research into appropriate waste-to-energy technologies for household, community and small-scale industrial uses, keeping energy and livelihood needs of the urban poor as a focus.

IS URBAN FORESTRY A SOLUTION TO THE ENERGY CRISIS OF SAHELIAN CITIES

The term 'urban forestry' indicates the simple fact that people purposefully plant and maintain trees amid their urban surroundings. However, urban forestry is by no means unique to developing countries; in fact, it is common for most regions of the world. The reasons for planting trees along roads, in public squares and private gardens are many:

- tangible reasons; such as food-production and fuelwood supplies
- intangible reasons; for recreation, to improve the quality of air, and to look attractive.

URBANIZATION AS A CHARACTERISTIC TRAIT OF DEVELOPING COUNTRIES

There is no doubt that increasing urbanization is a characteristic for most developing countries.

Growth rate in the amount of urbanized land in Africa (4.4) is almost twice as high as it is world wide.

People who abandon their rural lives in order to move to fast-growing cities, normally have high hopes for a better life;

food, employment opportunities and higher wages, energy etc. In most cases, though, this remains a dream, and a considerable number of people have insufficient access to these goods, services and advanced energy supplies such as gas, kerosene, and electricity. A large number of people remain dependent on subsistence food supplies, fuelwood or charcoal, and traditional cooking devices. Urban forestry can help to solve some of these problems.

An urban population's demand for urban forestry's tangible and intangible benefits varies in relation to the degree of development. Less developed urban societies are more dependent on individual, tangible benefits for their everyday existence, than are more highly developed urban communities. The latter in turn prefer social, intangible benefits.

This way, the consumption of fuelwood and/or charcoal by private urban households is closely linked with the level of development. This leads us to the question of which factors determine specific patterns of urban households' energy consumption.

FACTORS WHICH DETERMINE URBAN HOUSEHOLD ENERGY CONSUMPTION PATTERNS IN SAHELIAN COUNTRIES

Several factors, both physical and socio-economic, have been examined to determine private households' energy consumption patterns. Among these factors are:

- The availability of different energy sources,
- Different household sizes and cooking habits
- The available cooking technology
- Energy prices in relation to the households' mean income.

As a general rule it can be stated that - not very surprisingly - the consumption of fuelwood and charcoal is directly related to its availability. Therefore forest energy consumption per person in cities is normally lower than it is in rural areas. Outside urban areas, wood in most cases is available as a 'free' supply. Forest growth, which determines the availability of fuelwood in Sahelian Countries, is directly related to the mean annual rainfall. Since fuelwood consumption is related to fuelwood availability, it too is directly related to the mean annual rainfall. Other factors have an influence on wood production: different land use patterns (e.g. settlement structures, agriculture, agro-forestry etc.); the type of forests (with respect to their growth); the steepness of the landscape etc.

An important factor for urban citizens' preference for one energy source or another certainly is the price they have to pay on the market. Figure 2 compares the prices for electricity, kerosene, butane, fuelwood, and charcoal in Senegal. Not surprisingly, the latter two are by far the cheapest energy sources, while electricity is the most expensive.

While electricity may be most people's choice as a convenient source of energy, it requires a considerably developed and properly maintained public infrastructure, which - in most cases -is either not available, or too expensive for the average household. Many people have to rely on kerosene, and butane as alternatives to fuelwood and charcoal.

Since energy consumption patterns are closely linked to the level of development, one should expect that with the average household's income rising, the use of fuelwood and charcoal would be immediately reduced. In reality, this is usually not true as initially most people do not shift to more convenient (though more expensive) sources of energy; instead, they simply use more energy.

This behaviour becomes more evident where large price differences exist between 'modern' expensive energy sources and traditionally cheap energy sources, such as fuel-wood and charcoal. Besides, no new equipment has to be bought. Exploitation of forest resources will be the inevitable consequence.

Only when the average household's income increases so far that buying a new stove, gas burner or electric installation becomes a realistic option, people will shift to gas or electricity. The use of fuelwood and charcoal will then drop quite sharply, and will remain stabilized at a comparatively low level afterwards.

URBAN FORESTRY'S CONTRIBUTION TO THE SOLUTION OF SAHELIAN ENERGY PROBLEMS

Having addressed some of the prime reasons for urban households' energy consumption, we have to ask what conclusions may be drawn and how urban forestry's future role in Sahelian cities may be defined.

As urbanization continues, the demand for energy will continue to increases rapidly, due to the sheer number of inhabitants, especially if their standards of living improve. Successful development will - in the long run -induce energy substitution and will thus reduce the pressure on forests, not only in the vicinity of cities but also in rural supply areas.

However, many poverty-stricken people are likely to remain dependent on fuelwood and charcoal as a prime source of energy over the next decade. As has been stated in the previous section, the process of changing the energy consumption pattern will start with a significantly increased demand for traditional, 'cheap' energy sources. Exploitation of urban forests and of woodlands in the vicinity of major Sahelian cities, as well as in some rural supplier areas, is a likely outcome and may aggravate the already grave erosion and desertification problems, besides jeopardizing urban forestry's intangible benefits.

Strategies to limit this energy and environmental crisis comprise an increase in forest production as well as a reduction of fuel-wood/charcoal consumption. Looked at in the short term, dissemination of energy-saving technology might be the best bet, as it helps people to use forest energy both effectively and efficiently and has an immediate impact. However, it is also clear that increasing the Sahelian forest production will remain an important and justified effort. This has been tried with varying success through large-scale afforestation, social/ community forestry measures, and natural forests management. In the field of urban forestry very little information on its potential for fuelwood supply is available. However, one limiting factor is certainly the availability of suitable land. A rough estimate of the magnitude of a possible impact is made on the basis of the potential for forest production: Figure 3 shows the wood production of the fast growing eucalyptus tree compared with natural tree savannah for different amounts of rainfall.

If we take eucalyptus as the more productive alternative (although in reality certain species which respond better in urban areas would be chosen) it is possible to compare the theoretical wood production with the energy consumption in different Sahelian cities.

For each of the five major Sahelian cities noted in Figure, if they have at their - disposal an urban plantation of 300 hectares (corresponding to 187,500 trees), it can be seen that for the rainfall available in those regions, only 200 to 1500 citizens could be sustainably supplied with fuelwood.

These findings lead to the following conclusions:

Urban forestry is important if one considers the many positive impacts, but alone it cannot significantly contribute to mitigate the energy crises in Sahelian cities. The only possible option to improve the situation is to apply a holistic approach. This comprises such elements as support of different forest production systems, dissemination of fuelwood saving devices, fuelwood substitution, awareness creation and a favourable political and institutional framework. Such an approach has to include social, cultural, economic and technical aspects. It has to be flexible and adapted to local conditions. As part of such an approach, urban forestry can play an important role.

ELECTRICITY FOR LOW-POWER APPLICATIONS MICRO SOLAR LANTERNS FOR RURAL COMMUNITIES

A solar lantern has been developed for use in rural Kenyan homes using sunlight to charge a battery during the day, which then powers a lamp after dark. This project has been developed by IT Consultants, together with project partners International Development Enterprises, in Kenya. Using a combination of highly technical production methods and participatory market surveys, IT Consultants hope to come up with a product that is tailor-made for Kenyan customers.

Market Research

In order to develop the lantern effectively, the project team decided to carry out market research in two stages:

- The first stage (already complete) set out to determine which aspects of existing solar lantern designs were favourable to potential customers.
- The second stage will use working prototypes (designed with the results of the first survey clearly in mind) to gauge customers' reactions to the new design.

Market research, which is being carried out in several countries in the South is geared towards peoples' lifestyles and cultural backgrounds. Bearing this in mind, a number of studies have been carried out in Kenya by project partners Energy Alternatives Africa, which have attempted to identify which aspects of existing solar lantern designs are favoured by potential customers.

During the lantern survey, focus groups were used to measure reactions to particular design details. The ideas generated by these focus groups were recorded on paper and were also used to stimulate more general discussions where participants could air their viewpoints, and which often contained interesting and unexpected information.

- Survey groups yielded more accurate results when they were not dominated by a few knowledgeable individuals. Under these circumstances, everyone felt comfortable about having their say.
- The location and time of day for group surveys was important. Holding a survey after dark might at first seem to be the best way of demonstrating lanterns and their important characteristics to customers. In practice the team found that in rural areas with no street lighting, finding participants (women in particular) who were willing to travel from home at that time was more difficult!
- The team discovered that although participants drawn from slightly higher income groups are easier to access and interview, their spending patterns can differ from

those of customers from the lower income households at whom the new products might be targetted.

- It is often difficult to collect accurate information about how much a customer is willing to pay for a new product, especially in communities where bargaining is part of everyday business. The question 'how much would you pay?' when asked directly is more likely to lead to bargaining rather than to yield any useful information! Asking instead how much the participants friend or relative might pay helps to depersonalize the question and avoids a temptation to haggle.

As a result of the initial focus group study the project team were able to come up with a concise description of the 'ideal lantern'. The most important features were identified were as follows:

Service Characteristics

- the maximum price of the lantern should be no more than $75
- the lantern should provide light for up to 4 hours each evening
- customers should have access to affordable and readily available spares
- customers expect an overall lifetime of the lantern of 6 years
- customers expect a 12 months warranty for the product

Design Characteristics

- the lantern should give a 360 degree spread of light
- the bulb enclosure should allow maximum transmission of light with minimum dispersion
- the carry handle should be sturdy and comfortable
- the preferred choice of bulb was (5w cfl type)
- the lamp should be portable and weigh no more than 2.5 kg
- the lantern should be stable with a good base

Extra features for which potential customers expressed a need were:

- an indicator to show that lamp is charging
- a warning light to show that the lamp is about to switch off when the battery is low
- a power socket to allow a small radio to be connected to the unit.

The findings of this initial survey were used to produce a design, and as a result, the team have produced a new design for an injection moulded lamp which incorporates all of these features.

Rapid Prototyping

It is vital during the development of any new product to show customers a sample and to listen to their ideas about it. This is especially important if producing large quantities involves a lot of money to set up tooling and machinery. Until a few years ago designers could only produce 'block models' of new products of this type. These were constructed by hand from wood and plastic, and although they had the appearance of the final product they could not normally demonstrate how the new product would work. Today, computer-aided design (CAD), combined with rapid prototyping techniques and soft-tooling allows the designer to make a single copy of their new design in a matter of hours. This technique allows the designer to find out very quickly how his design will work, and more importantly, allows potential customers to give their opinions, all before any significant costs have been incurred.

In developing the solar lantern, the project team have been able to use facilities at Coventry University, UK, to produce a rapid prototype of the new lantern. This will in turn be used to produce a small batch of sample lanterns that will be tested in sample households in Kenya. This phase of research, used to measure customers' reactions to the new lantern design, has been designed by IT Consultants' project partners International Development Enterprises, working closely with ITDG staff in Kenya. Fully working test lanterns will be distributed to sample households where they will be used for a period of one week.

Project researchers will then visit households and use a simple questionnaire to measure customers reactions.

If all goes well and the sample households feel positive about the lanterns, the next stage will be to pay for someone to make the tools for the injection-moulded components. It is likely that these will be produced in UK and then shipped to Kenya where the moulding and assembly of the lanterns will be carried out by locally appointed companies.

The ultimate success of any new product can be measured through sales figures. The first estimates show a potential market for over one million units in Kenya alone, if cost can be kept low. IT Consultants expect to use the experience gained in Kenya to assess the size of the demand for solar lanterns in other countries of the South. It will also be assessing the success of the international technological collaboration between IT Consultants in the UK and International Development Enterprises in Kenya.

THE BICYCLE WHEEL WATER POWERED BATTERY CHARGER

Throughout East Africa, old car batteries find a second home as a source of electric power for those living off the grid. Radios and small black and white television sets are the most common appliances powered by the battery, though some domestic electric lighting competes with the more popular kerosene lamp.

Batteries typically last ten days to two weeks in between charges, and it can cost as much as US$ 1.50 to be recharged by a village-based entrepreneur operating a charging station off the national grid. This regular top-up can entail a trek of many kilometres over rugged countryside, often carrying the heavy battery by bicycle or even balanced on a person's head. Some of the roughest and most inaccessible terrain is in mountainous regions, where plentiful sources of water are available which can be tapped to drive this simple battery charger.

A Chinese-made bicycle headlamp generator, of a type widely available throughout Africa, has an output of 12 Volts and 5.5 watts. It struck me that this was just under one amp, which was sufficient to trickle charge a car battery to full power over a ten to twelve hour period.

The unit cost breakdown (in US$ equivalent):

One almost-scrap bicycle:	11.30
Chinese-made bicycle headlamp and generator set:	6.55
Scrap iron, wires, clips and diodes:	7.40
Total Materials Cost:	25.5

Upon testing the brush-less 'generator' against the rotating tyre of my bicycle, I found that it produced an alternating current. A diode bridge was required to rectify the current to direct current in order to charge a car battery (Figure 1). Soldering four diodes together in the right direction is simple enough and works well, though at the expense of about 0.6 of a volt due to resistance within the diode; this can be compensated by increasing the speed of the waterwheel slightly.

Figure 1

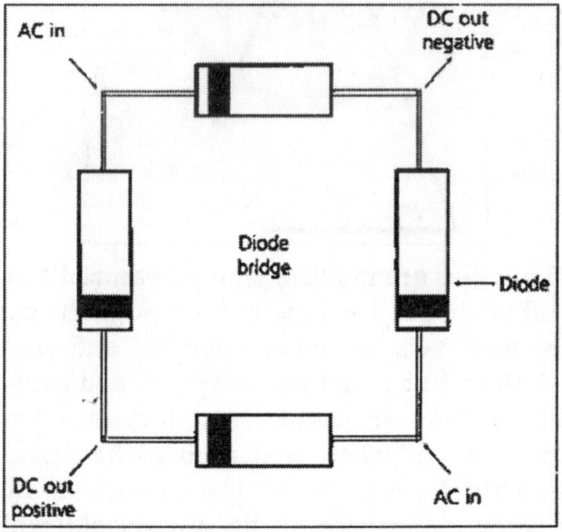

An old, well-used bicycle was purchased from a willing vendor and the two wheels, complete with threadbare tyrės, were mounted on a shared axle separated by six inches. The axle and wheels are supported by a simple double A-frame, with a two foot pipe extending up from one side acting as a mount for the generator which runs (as it should) on the side-wall of the tyre (not the rim). The frame is made of welded scrap angle iron (Figure 2).

Figure 2

The 16 paddles are made of thin galvanized iron salvaged from old roofing sheets. They are cut to bridge the gap between the two bicycle wheels, which are parallel and six inches (15 cm) apart at their hubs, and are simply folded over spokes on each wheel. If you can visualize a bicycle wheel's spoke arrangement, you will know that each spoke crosses another at about midpoint, forming an 'X'. The paddles are folded along a single spoke on either side of this cross-spoke which acts to centralize the paddles- preventing movement towards either the rim or the hub. Each paddle is eight inches (20 cm) long and extends to within four inches (10 cm) of the 21 inch (52 cm) diameter bicycle rim (Figure 3).

Figure 3

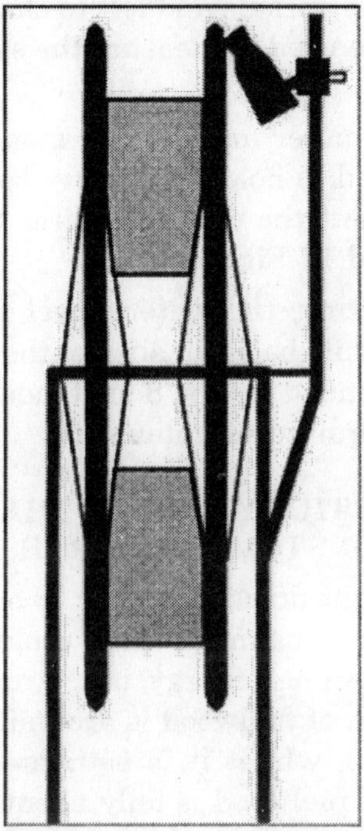

A stream of water from a two-inch pipe placed three to four feet above the paddlewheel serves to drive the unit.

Figure 4

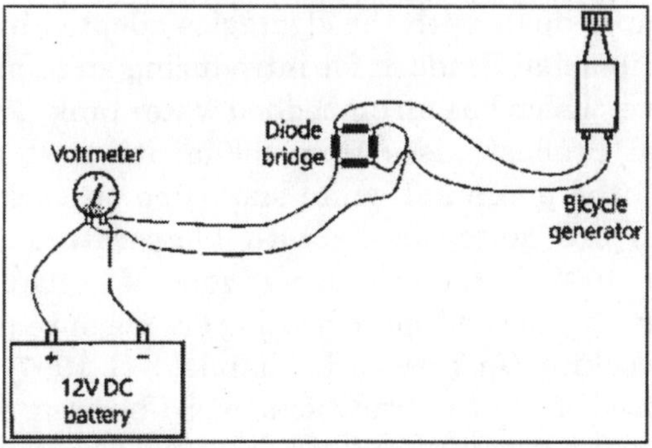

The paddlewheel must rotate at close to 100 rpm in order to spin the generator on the side wall of one tyre with enough speed to generate 13.5 volts. The battery is simply connected to the appropriate terminals and left to charge. A voltmeter or a hygrometer can be used to measure the state of charge of the battery.

An in-line voltmeter (optional) is mounted in the lid of a wooden box designed to hold the battery being charged. Large alligator clips connect the charging wires to the battery. The wiring is illustrated in Figure 4.

At time of writing, the water-wheel generator has been used on an almost daily basis for six months without a problem. It can be left unattended for the 8 to 10 hours charging period, and has not yet required maintenance.

COMMUNITY PARTICIPATION IN THE DEVELOPMENT OF AN IMPROVED STOVE IN NORTH INDIA

In rural India, the domestic sector accounts for nearly 75% of the total energy consumed; with cooking accounting for almost 90% of household energy use. Annually, the present level of consumption of fuelwood is around 200 million tonnes (Sinha C. S. et. al.), whilst it is estimated that the annual sustainable yield of fuelwood is only about 86 million tonnes. Thus, a wide gap exists between demand and sustainable supply. At the same time, traditional stoves have serious effects on the environment and lead to high levels of indoor air pollution, which affect the health of women and children.

This paper deals with the strategies adopted in the Solan district of Himachal Pradesh for introducing an improved mud cooking stove, which has an embedded water tank. The altitude of Himachal Pradesh rises from 900m, its most prominent landmark is the perennial white snow-line on various peaks. The winters are quite severe with temperature sometimes dropping to -400C in the cold desert zone of Lahaul and Spiti districts. Thus, water and space heating are major requirements, apart from cooking (Aggarwal R, Chandel S S, 1997). The most commonly used stoves are traditional wood-burning mud stoves with efficiencies ranging from 8% to 10%. The common fuel species are pinus roxburghii, and Banj (Quercus), as well as a few other less common shrubs.

The traditional cooking stove was modified to meet the need for hot water expressed by the local community.

Community Inputs

Initial results were obtained from a water-heating tank attached to a stove, which had been developed by a local housewife. An inbuilt water tank was -placed at the bottom of the firebox with connections to the bathroom. This design was modified to develop the TERI model of improved stove for the village. The new model was installed in a test household and monitored for one month. The design was modified several times after discussions with the households. The shape and design of the kitchen, and user preferences were considered, so that the stove design could be modified slightly to suit individual households.

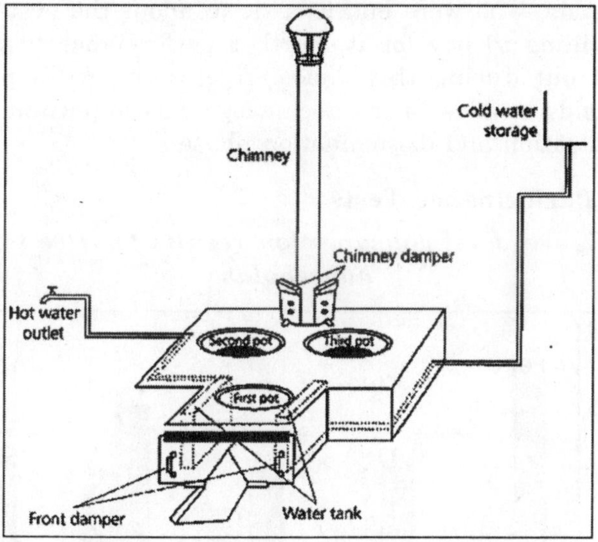

Stove characteristics : Two embedded water tanks (combined capacity 15 litres) one on each side of the fire box. These two tanks are connected to each other, they get their cold water supply from an overhead tank and the hot water can be drawn from a tap provided in the kitchen or the bathroom. Heat is conducted to the water tanks from the firebox, and hot water is available after igniting the stove. The tanks are covered with mud on the outside to reduce heat loss. Usually, a steam pipe is fitted to the overhead tank to vent the system. A chimney channels the smoke outside. A front damper maintains the temperature in the stove and water pipes. Figure 1: Schematic drawing of improved stove.

Community Co-operation

Important factors in this project were:

- The involvement of the test household in the project
- Their willingness to cope with repeated instructions and construction
- The co-operation of household members in providing valuable ideas for modification

The next step was to install the stove in five sample households, who were enthusiastic to adopt the system and were willing to pay for it. Further performance tests were carried out during this phase. Again, the willingness of households to adopt a test technology was important in this demonstration and dissemination phase.

Stove Performance Tests

Figure: Fuel consumption results for five test households

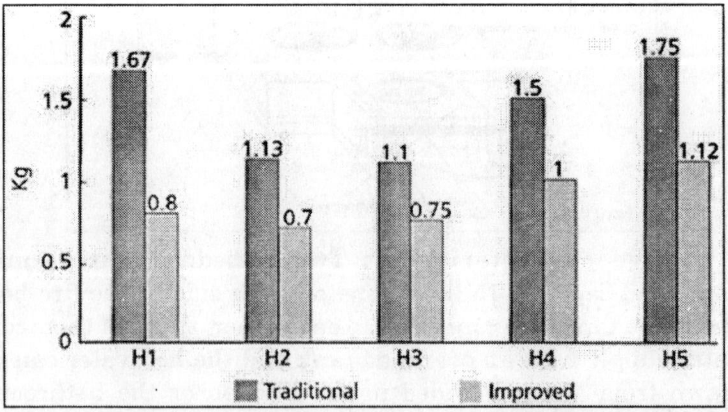

Two days were spent in making and installing the stove in each of the five households. During this time, household members were given detailed information on how the stove worked. A second visit was made after 4-5 days to show households how to use the stove. Ten to twenty days after installation, tests were carried out using improved and traditional models to compare fuel consumption and cooking time (tests were repeated); results showed that households saved both fuel and cooking time; on average, about 45%.

At the same time, users were asked what they thought about the improved model compared to their traditional stoves.

Figure: Cooking time results (traditional and improved cooking stoves)

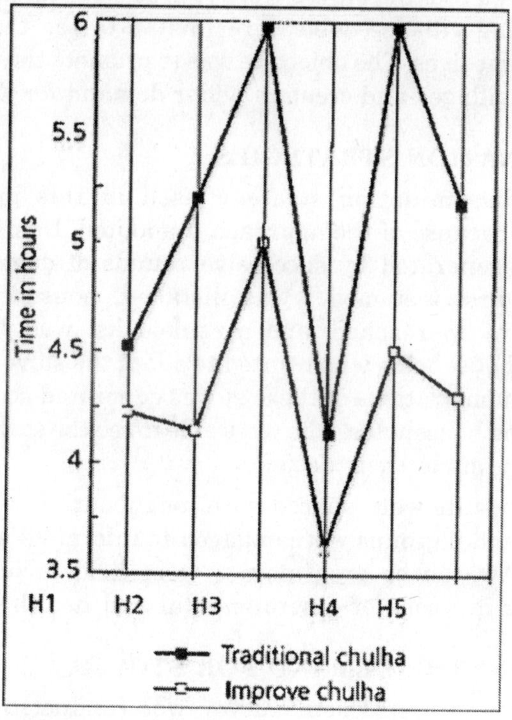

TECHNOLOGY PROMOTION

The improved stove models installed in the five households were used to demonstrate the technology to the local communities, and to provide data for technical performance assessment. To convince the village community about the benefits of the technology, the Mahila Mandal was asked to disseminate information as it was believed to have closer contact and more credibility with the community. (The Mahila Mandal is a social group, organized by women at village level for various social, economic and religious activities.)

At the same time, awareness programmes for village women were created dealing with: fuel-wood consumption; over-exploitation of resources, deforestation, health hazards from indoor pollution, and the benefits of improved cooking stove technology. The test results from the performance surveys of the improved cooking stoves were shared with communities in neighbouring villages, who were invited to see the installed models themselves. The objective was to promote the technology in several villages and create a wider demand for the devices.

DISSEMINATION STRATEGIES

The dissemination strategy used in this project was successful because of the approach it adopted. Demand for the stove was generated by successive rounds of demonstration. After the first test model was installed, households in the village were approached, and a ranked list was prepared of volunteer households who wanted to adopt the stove in the first phase. Demonstrations of these stoves generated acceptance in surrounding households, who were able to see the stoves working before they made their decision.

Test results were shared with neighbouring villages, and Mahila Mandal groups were engaged to inform villagers of the benefits. Awareness programmes were held to promote the technology in terms of environmental and health benefits.

ASSESSING THE DEMAND FOR STOVES

A demand assessment survey was conducted in several villages. The residents of nearby villages could visit the households and inspect the stoves for themselves and talk to the test households before making up their minds. About 20 households adopted the technology at this stage. When more stoves were installed, additional demand was generated.

Flexibility in Stove Design

Another aspect of acceptance was the flexibility of the improved cooking stove. Households could have extra features installed or the system modified according to their choice if they were willing to pay the additional cost.

In all 86 rural families from the 15 small villages benefited, improving their kitchens by giving them efficient and user-friendly cooking and water-heating devices. The most visible benefit was the availability of hot water without making extra work, and there were other gains, including fuel and time saved, and reduced smoke emissions during the cooking. The technology was well received, leading to demand from other households. It was noticed that households using the improved stove benefited by having more free time and being less tired. This has a direct impact on the time and care they can devote to their children, and the quality of their kitchen environment. It is also expected to provide an annual saving of 260 tonnes of biomass in the region.

Future Plans

The project has now ended and the programme has been handed on to a local NGO. Stoves are now being installed by local entrepreneurs, who have found that the design is well-accepted and considered desirable by the local community. Consequently, no further modifications are planned for the stove, although increasing its overall efficiency might be possible. It is currently priced at Rs 500-600 (Rs70 = £1UK) depending on the amount of pipe-work required, and compares favourably with the cost of an equivalent metal stove.

Gas Turbine

A gas turbine, also called a combustion turbine, is a rotary engine that extracts energy from a flow of combustion gas. It has an upstream compressor coupled to a downstream turbine, and a combustion chamber in-between. (Gas turbine may also refer to just the turbine element.)

Energy is added to the gas stream in the combustor, where air is mixed with fuel and ignited. Combustion increases the temperature, velocity and volume of the gas flow. This is directed through a nozzle over the turbine's blades, spinning the turbine and powering the compressor.

Energy is extracted in the form of shaft power, compressed air and thrust, in any combination, and used to power aircraft, trains, ships, generators, and even tanks.

History

- 150: Hero's Engine (*aeolipile*) - apparently Hero's steam engine was taken to be no more than a toy, and thus its full potential not realized for centuries.
- 1500: The "Chimney Jack" was drawn by Leonardo da Vinci which was turning a roasting spit. Hot air from a fire rose through a series of fans which connect and turn the roasting spit.
- 1551: Taqi al-Din invented a steam turbine, which he used to power a self-rotating spit.
- 1629: Jets of steam rotated a turbine that then rotated driven machinery allowed a stamping mill to be developed by Giovanni Branca.
- 1678: Ferdinand Verbeist built a model carriage relying on a steam jet for power.
- 1791: A patent was given to John Barber, an Englishman, for the first true gas turbine. His invention had most of the elements present in the modern day gas turbines. The turbine was designed to power a horseless carriage.
- 1872: The first true gas turbine engine was designed by Dr Franz Stolze, but the engine never ran under its own power.
- 1894: Sir Charles Parsons patented the idea of propelling a ship with a steam turbine, and built a demonstration vessel (the Turbinia). This principle of propulsion is still of some use.
- 1895: Three 4-ton 100 kW Parsons radial flow generators were installed in Cambridge Power Station, and used to power the first electric street lighting scheme in the city.

- 1903: A Norwegian, Ægidius Elling, was able to build the first gas turbine that was able to produce more power than needed to run its own components, which was considered an achievement in a time when knowledge about aerodynamics was limited. Using rotary compressors and turbines it produced 11 hp (massive for those days). His work was later used by Sir Frank Whittle.
- 1913: Nikola Tesla patents the Tesla turbine based on the Boundary layer effect.
- 1914: Application for a gas turbine engine filed by Charles Curtis.
- 1918: One of the leading gas turbine manufacturers of today, General Electric, started their gas turbine division.
- 1920: The practical theory of gas flow through passages was developed into the more formal (and applicable to turbines) theory of gas flow past airfoils by Dr A. A. Griffith.
- 1930: Sir Frank Whittle patented the design for a gas turbine for jet propulsion. His work on gas propulsion relied on the work from all those who had previously worked in the same field and he has himself stated that his invention would be hard to achieve without the works of Ægidius Elling. The first successful use of his engine was in April 1937.
- 1934: Raúl Pateras de Pescara patented the free-piston engine as a gas generator for gas turbines.
- 1936: Hans von Ohain and Max Hahn in Germany developed their own patented engine design at the same time that Sir Frank Whittle was developing his design in England.

Theory of Operation

Gas turbines are described thermodynamically by the Brayton cycle, in which air is compressed isentropically, combustion occurs at constant pressure, and expansion over the turbine occurs isentropically back to the starting pressure.

In practice, friction, and turbulence cause:

1. non-isentropic compression: for a given overall pressure ratio, the compressor delivery temperature is higher than ideal.
2. non-isentropic expansion: although the turbine temperature drop necessary to drive the compressor is unaffected, the associated pressure ratio is greater, which decreases the expansion available to provide useful work.
3. pressure losses in the air intake, combustor and exhaust: reduces the expansion available to provide useful work.

As with all cyclic heat engines, higher combustion temperature means greater efficiency. The limiting factor is the ability of the steel, nickel, ceramic, or other materials that make up the engine to withstand heat and pressure. Considerable engineering goes into keeping the turbine parts cool. Most turbines also try to recover exhaust heat, which otherwise is wasted energy. Recuperators are heat exchangers that pass exhaust heat to the compressed air, prior to combustion. Combined cycle designs pass waste heat to steam turbine systems. And combined heat and power (co-generation) uses waste heat for hot water production.

Mechanically, gas turbines can be considerably less complex than internal combustion piston engines. Simple turbines might have one moving part: the shaft/compressor/turbine/alternative-rotor assembly (see image above), not counting the fuel system.

More sophisticated turbines (such as those found in modern jet engines) may have multiple shafts (spools), hundreds of turbine blades, movable stator blades, and a vast system of complex piping, combustors and heat exchangers.

As a general rule, the smaller the engine the higher the rotation rate of the shaft(s) needs to be to maintain top speed. Turbine blade top speed determines the maximum pressure that can be gained,this produces the maximum power possible independent of the size of the engine. Jet engines operate around 10,000 rpm and micro turbines around 100,000 rpm.

Thrust bearings and journal bearings are a critical part of design. Traditionally, they have been hydrodynamic oil bearings, or oil-cooled ball bearings. These bearings are being surpassed by foil bearings, which have been successfully used in micro turbines and auxiliary power units.

TYPES OF GAS TURBINES

Aeroderivatives and Jet Engines

Airbreathing jet engines are gas turbines optimized to produce thrust from the exhaust gases, or from ducted fans connected to the gas turbines. Jet engines that produce thrust primarily from the direct impulse of exhaust gases are often called turbojets, whereas those that generate most of their thrust from the action of a ducted fan are often called turbofans or (rarely) fan-jets.

Gas turbines are also used in many liquid propellant rockets, the gas turbines are used to power a turbopump to permit the use of lightweight, low pressure tanks, which saves considerable dry mass.

Aeroderivatives are also used in electrical power generation due to their ability to startup, shut down, and handle load changes more quickly than industrial machines. They are also used in the marine industry to reduce weight. The GE LM2500 and LM6000 are two common models of this type of machine.

Amateur Gas Turbines

Increasing numbers of gas turbines are being used or even constructed by amateurs. In its most straightforward form, these are commercial turbines acquired through military surplus or scrapyard sales, then operated for display as part of the hobby of engine collecting. In its most extreme form, amateurs have even rebuilt engines beyond professional repair and then used them to compete for the Land Speed Record.

The simplest form of self-constructed gas turbine recycles the turbine wheel and compressor from an automotive turbocharger. A single separate combustion chamber is fabricated and plumbed between the compressor and turbine.

More sophisticated turbojets are also built, where their thrust and light weight are sufficient to power large model aircraft. The Schreckling design constructs the entire engine from raw materials, including the fabrication of a centrifugal compressor wheel from plywood, epoxy and wrapped carbon fibre strands.

Like many technology based hobbies, they tend to give rise to manufacturing businesses over time. Several small companies now manufacture small turbines and parts for the amateur. Most turbojet-powered model aircraft are now using these commercial and semi-commercial microturbines, rather than a Schreckling-like home-build.

Auxiliary Power Units

Auxiliary power units (APUs) are small gas turbines designed for auxiliary power of larger machines, such as those inside an aircraft. They supply compressed air for aircraft ventilation (with an appropriate compressor design), start-up power for larger jet engines, and electrical and hydraulic power.

INDUSTRIAL GAS TURBINES FOR ELECTRICAL GENERATION

Industrial gas turbines differ from aeroderivatave in that the frames, bearings, and blading is of heavier construction. Industrial gas turbines range in size from truck-mounted mobile plants to enormous, complex systems. They can be particularly efficient—up to 60%—when waste heat from the gas turbine is recovered by a heat recovery steam generator to power a conventional steam turbine in a combined cycle configuration. They can also be run in a cogeneration configuration: the exhaust is used for space or water heating, or drives an absorption chiller for cooling or refrigeration. A cogeneration configuration can be over 90% efficient. The power turbines in the largest industrial gas turbines operate at 3,000 or 3,600 rpm to match the AC power grid frequency and to avoid the need for a reduction gearbox. Such engines require a dedicated enclosure, both to protect the engine from the elements and the operators from the noise.

Simple cycle gas turbines in the power industry require smaller capital investment than either coal or nuclear power plants and can be scaled to generate small or large amounts of power. Also, the actual construction process can take as little as several weeks to a few months, compared to years for base load power plants.

Their other main advantage is the ability to be turned on and off within minutes, supplying power during peak demand. Because they are less efficient than combined cycle plants, they are usually used as peaking power plants, which operate anywhere from several hours per day to a couple dozen hours per year, depending on the electricity demand and the generating capacity of the region. In areas with a shortage of base load and load following power plant capacity, a gas turbine power plant may regularly operate during most hours of the day and even into the evening. A typical large simple cycle gas turbine may produce 100 to 300 megawatts of power and have 35–40% thermal efficiency. The most efficient turbines have reached 46% efficiency.

COMPRESSED AIR ENERGY STORAGE

One modern development seeks to improve efficiency in another way, by separating the compressor and the turbine with a compressed air store. In a conventional turbine, up to half the generated power is used driving the compressor. In a compressed air energy storage configuration, power, perhaps from a wind farm or bought on the open market at a time of low demand and low price, is used to drive the compressor, and the compressed air released to operate the turbine when required.

Turboshaft Engines

Turboshaft engines are often used to drive compression trains (for example in gas pumping stations or natural gas liquefaction plants) and are used to power almost all modern helicopters. The first shaft bears the compressor and the high speed turbine (often referred to as "Gas Generator" or "N1"), while the second shaft bears the low speed turbine (or "Power Turbine" or "N2"). This arrangement is used to increase speed and power output flexibility.

Radial Gas Turbines

1963, Norway, Jan Mowill initiated the development at Kongsberg Våpenfabrikk. Various successors have made good progress in the refinement of this mechanism. Owing to a configuration that keeps heat away from certain bearings the durability of the machine is improved while the radial turbine is well matched in speed requirement

Scale Jet Engines

Also known as miniature gas turbines or micro-jets.

Many model engineers relish the challenge of re-creating the grand engineering feats of today as tiny working models. Naturally, the idea of re-creating a powerful engine such as the jet, fascinated hobbyists since the very first full size engines were powered up by Hans von Ohain and Frank Whittle back in the 1930s.

Recreating machines such as engines to a different scale is not easy. Because of the square-cube law, the behaviour of many machines does not always scale up or down at the same rate as the machine's size (and often not even in a linear way), usually at best causing a dramatic loss of power or efficiency, and at worst causing them not to work at all. An automobile engine, for example, will not work if reproduced in the same shape at the size of a human hand.

With this in mind the pioneer of modern Micro-Jets, Kurt Schreckling, produced one of the world's first Micro-Turbines, the FD3/67. This engine can produce up to 22 newtons of thrust, and can be built by most mechanically minded people with basic engineering tools, such as a metal lathe. Its radial compressor, which is cold, is small and the hot axial turbine is large experiencing more centrifugal forces, meaning that this design is limited by Mach number. Guiding vanes are used to hold the starter, after the compressor impeller and before the turbine. No bypass within the engine is used.

Microturbines

Also known as:

- Turbo alternators

- MicroTurbine (registered trademark of Capstone Turbine Corporation)
- Turbogenerator (registered tradename of Honeywell Power Systems, Inc.)

Microturbines are becoming widespread for distributed power and combined heat and power applications. They are one of the most promising technologies for powering hybrid electric vehicles. They range from hand held units producing less than a kilowatt, to commercial sized systems that produce tens or hundreds of kilowatts.

Part of their success is due to advances in electronics, which allows unattended operation and interfacing with the commercial power grid. Electronic power switching technology eliminates the need for the generator to be synchronized with the power grid. This allows the generator to be integrated with the turbine shaft, and to double as the starter motor.

Microturbine systems have many advantages over reciprocating engine generators, such as higher power density (with respect to footprint and weight), extremely low emissions and few, or just one, moving part. Those designed with foil bearings and air-cooling operate without oil, coolants or other hazardous materials. Microturbines also have the advantage of having the majority of their waste heat contained in their relatively high temperature exhaust, whereas the waste heat of recriprocating engines is split between its exhaust and cooling system. However, reciprocating engine generators are quicker to respond to changes in output power requirement and are usually slightly more efficient, although the efficiency of microturbines is increasing. Microturbines also lose more efficiency at low power levels than reciprocating engines.

They accept most commercial fuels, such as gasoline, natural gas, propane, diesel, and kerosene as well as renewable fuels such as E85, biodiesel and biogas.

Microturbine designs usually consist of a single stage radial compressor, a single stage radial turbine and a recuperator. Recuperators are difficult to design and manufacture because they operate under high pressure and temperature differentials.

Exhaust heat can be used for water heating, space heating, drying processes or absorption chillers, which create cold for air conditioning from heat energy instead of electric energy.

Typical microturbine efficiencies are 25 to 35%. When in a combined heat and power cogeneration system, efficiencies of greater than 80% are commonly achieved.

MIT started its millimeter size turbine engine project in the middle of the 1990s when Professor of Aeronautics and Astronautics Alan H. Epstein considered the possibility of creating a personal turbine which will be able to meet all the demands of a modern person's electrical needs, just like a large turbine can meet the electricity demands of a small city. Problems have occurred with heat dissipation in these new microturbines. According to Professor Epstein current commercial Li-ion rechargeable batteries deliver about 120-150 Wh/kg. MIT's millimeter size turbine will deliver 500-700 Wh/kg in the near term, rising to 1200-1500 Wh/kg in the longer term.

External Combustion

Most gas turbines are internal combustion engines but it is also possible to build an external combustion gas turbine which is, effectively, a turbine version of a hot air engine.

External combustion has been used for the purpose of using pulverized coal or finely ground biomass (such as sawdust) as a fuel. External combustion gas has been used both directly and indirectly. In the direct system, the combustion products travel through the power turbine. In the indirect system, a heat exchanger is used and clean air travels through the power turbine. The thermal efficiency is lower in the indirect type of external combustion, however the blades are not subjected to combustion products.

GAS TURBINES IN VEHICLES

Gas turbines are used on ships, locomotives, helicopters, and in tanks. A number of experiments have been conducted with gas turbine powered automobiles. In 1950, designer F.R. Bell and Chief Engineer Maurice Wilks from British car manufacturers Rover unveiled the first car powered with a gas turbine engine. The two-seater JET1 had the engine positioned behind the seats, air intake grilles on either side of the car, and exhaust outlets on the top of the tail. During tests, the car reached top speeds of 140 km/h, at a turbine speed of 50,000 rpm. The car ran on petrol, paraffin or diesel oil, but fuel consumption problems proved insurmountable for a production car. It is currently on display at the London Science Museum.

Rover and the British Racing Motors (BRM) Formula One team joined forces to produce the Rover-BRM, a gas turbine powered coupe, which entered the 1963 24 Hours of Le Mans, driven by Graham Hill and Richie Ginther. It averaged 107.8 mph (173 km/h) and had a top speed of 142 mph (229 km/h). American Ray Heppenstall joined Howmet Corporation and McKee Engineering together to develop their own gas turbine sports car in 1968, the Howmet TX, which ran several American and European events, including two wins, and also participated in the 1968 24 Hours of Le Mans. The cars used Continental gas turbines, which eventually set six FIA land speed records for turbine-powered cars.

For open wheel racing, 1967's revolutionary *STP Oil Treatment Special* four-wheel drive turbine-powered special fielded by racing and entrepreneurial legend Andy Granatelli and driven by Parnelli Jones nearly won the Indianapolis 500; the STP Pratt & Whitney powered turbine car was almost a lap ahead of the second place car when a gearbox bearing failed just three laps from the finish line. In 1971 Lotus principal Colin Chapman introduced the Lotus 56B F1 car, powered by a Pratt & Whitney gas turbine. Chapman had a reputation of building radical championship-winning cars, but had to abandon the project because there were too many problems with turbo lag.

The original General Motors Firebird was a series of concept cars developed for the 1953, 1956 and 1959 Motorama auto shows, powered by gas turbines.

American car manufacturer Chrysler demonstrated several prototype gas turbine-powered cars from the early 1950s through the early 1980s. Chrysler built fifty Chrysler Turbine Cars in

1963 and conducted the only consumer trial of gas turbine-powered cars. Their turbines employed unique rotating recuperator that significantly increased efficiency.

Japanese car manufacturer Toyota demonstrated several gas turbine powered prototype vehicles such as the Century gas turbine hybrid in 1975, the Sports 800 Gas Turbine Hybrid in 1977 and the GTV in 1985. No production vehicles were made.

The fictional Batmobile is often said to be powered by a gas turbine or a jet engine. In fact, in 1989s filmed *Batman*, the production department built a working turbine vehicle for the Batmobile prop. Its fuel capacity, however, was reportedly only enough for 15 seconds of use at a time.

In 1993 General Motors introduced the first commercial gas turbine powered hybrid vehicle—as a limited production run of the EV-1 series hybrid. A Williams International 40 kW turbine drove an alternator which powered the battery-electric powertrain. The turbine design included a recuperator. Later on in 2006 GM went into the EcoJet concept car project with Jay Leno.

The arrival of the Capstone Microturbine has led to several hybrid bus designs, starting with HEV-1 by AVS of Chattanooga, Tennessee in 1999, and closely followed by Ebus and ISE Research in California, and Designline in New Zealand. AVS turbine hybrids were plagued with reliability and quality control problems, resulting in liquidation of AVS in 2003. The most successful design by Designline is now operated in 5 cities in 6 countries, with over 30 buses in operation worldwide.

A key advantage of jets and turboprops for aeroplane propulsion - their superior performance at high altitude compared to piston engines, particularly naturally-aspirated ones - is irrelevant in automobile applications. Their power-to-weight advantage is far less important.

Gas turbines offer a high-powered engine in a very small and light package. However, they are not as responsive and efficient as small piston engines over the wide range of RPMs and powers needed in vehicle applications. In hybrids, gas turbines reduce the responsiveness problem, and the emergence of the continuously variable transmission may also help alleviate this.

Turbines have historically been more expensive to produce than piston engines, though this is partly because piston engines have been mass-produced in huge quantities for decades, while small gas turbine engines are rarities; but turbines are mass produced in the closely related form of the turbocharger.

The MTT Turbine SUPERBIKE appeared in 2000 (hence the designation of Y2K Superbike by MTT) and is the first production motorcycle powered by a turbine engine - specifically, a Rolls-Royce Allison model 250 turboshaft engine, producing about 283 kW (380 bhp). Speed-tested to 365 km/h or 227 mph (according to some stories, the testing team ran out of road during the test), it holds the Guinness World Records for most powerful production motorcycle and most expensive production motorcycle, with a price tag of US$185,000.

Several locomotive classes have been powered by gas turbines, the most recent incarnation being Bombardier's JetTrain. See gas turbine-electric locomotive for more information.

Tank Use

The first use of a gas turbine in an armoured fighting vehicle was in 1954 when a unit, PU2979, specifically developed for tanks by C. A. Parsons & Co., was installed and trialled in a British Conqueror tank. Since then, gas turbine engines have been used as auxiliary power units (APUs) in some tanks and as main powerplants in Soviet/Russian T-80s and U.S. M1 Abrams tanks, among others. They are lighter and smaller than diesels at the same sustained power output but the models installed to date are less fuel efficient than the equivalent diesel, especially at idle, requiring more fuel to achieve the same combat range. Successive models of M1 have addressed this problem with battery packs or secondary generators to power the tank's systems while stationary, saving fuel by reducing the need to idle the main turbine. T-80s can mount three large external fuel drums to extend their range. Russia has stopped production of the T-80 in favour of the diesel-powered T-90 (based on the T-72), while Ukraine has developed the diesel-powered T-80UD and T-84 with nearly the power of the gas-turbine tank.

A turbine is theoretically more reliable and easier to maintain than a piston engine, since it has a simpler construction with fewer moving parts but in practice turbine parts experience a higher wear rate due to their higher working speeds. The turbine blades are highly sensitive to dust and fine sand, so that in desert operations air filters have to be fitted and changed several times daily. An improperly fitted filter, or a bullet or shell fragment that punctures the filter can damage the engine. Piston engines also need well-maintained filters, but they are more resilient if the filter does fail.

Like most modern diesel engines used in tanks, gas turbines are usually multi-fuel engines.

Naval Use

Gas turbines are used in many naval vessels, where they are valued for their high power-to-weight ratio and their ships' resulting acceleration and ability to get underway quickly.

The first gas-turbine-powered naval vessel was the Royal Navy's Motor Gun Boat *MGB 2009* (formerly *MGB 509*) converted in 1947. Metrovick developed the "Beryl" engine equipping an existing F2/3 jet engine with a power turbine. As the test was successful, the Fast Patrol Boats *Bold Pioneer* and *Bold Pathfinder* built in 1953 were the first ships created specifically for gas turbine propulsion.

The first large, gas-turbine powered ships, were the Royal Navy's Type 81 (Tribal class) frigates, the first of which (HMS *Ashanti*) was commissioned in 1961.

The Germany Navy launched the first Köln class frigate in 1961 with 2 GTs from BBC in the worlds first combined diesel and gas propulsion system.

The Swedish Navy produced 6 Spica class torpedoboats between 1966 and 1967 powered by 3 Bristol Siddeley Proteus 1282, each delivering 4300 hp. They were later joined by 12 upgraded Norrköping class ships, still with the same engines. With their aft torpedo tubes replaced by antishipping missiles they served as missile boats until the last was retired in 2005.

The Finnish Navy issued two Turunmaa class corvettes, Turunmaa and Karjala, in 1968. They were equipped with one 16 000 shp Rolls-Royce Olympus TMB3 gas turbine and two Wärtsilä marine diesels for slower speeds. Before the waterjet-propulsion Helsinki class missile boats, they were the fastest vessels in the Finnish Navy; they regularly achieved 37 knot speeds, but they are known to have achieved 45 knots when the restriction mechanism of the turbine was geared off. The Turunmaas were paid off in 2002. *Karjala* is today a museum ship in Turku, and *Turunmaa* serves as a flotating machine shop and training ship for Satakunta Polytechnical College.

The next series of major naval vessels were the four Canadian Iroquois class helicopter carrying destroyers first commissioned in 1972. They used 2 ft-4 main propulsion engines, 2 ft-12 cruise engines and 3 Solar Saturn 750 kW generators.

The first U.S. gas-turbine powered ships were the U.S. Coast Guard's *Hamilton*-class High Endurance Cutters the first of which (USCGC *Hamilton*) commissioned in 1967. Since then, they have powered the U.S. Navy's *Perry*-class frigates, *Spruance*-class and *Arleigh Burke*-class destroyers, and *Ticonderoga*-class guided missile cruisers. USS *Makin Island*, a modified *Wasp*-class amphibious assault ship, is to be the Navy's first amphib powered by gas turbines. The marine gas turbine operates in a more corrosive atmosphere due to presence of sea salt in air and fuel and use of cheaper fuels.

Commercial Use

There have been a number of experiments in which gas turbines were used to power seagoing commercial vessels. The earliest of these experiments may have been the oil tanker "Auris" (Anglo Saxon Petroleum) - circa 1949.

The United States Maritime Commission were looking for options to update WWII Liberty ships and heavy duty gas turbines were one of those selected. In 1956 The "John Sergeant" was lenghened and installed with a General Electric 6600 SHP

HD gas turbine, reduction gearing and a variable pitch propeller. It operated for 9700 hours using residual fuel for 7000 hours. The success of this trial opened the way for more development by GE on the use of HD gas turbines for marine use with heavy fuels. The "John Sergeant" was scrapped in 1972 at Portsmouth PA.

Between 1970 and 1982, Seatrain Container Lines operated a scheduled container service across the North Atlantic with four 26,000 tonne dwt. container ships. Those ships were powered by twin Pratt & Whitney gas turbines of the FT 4 series. The four ships in the class were named "Euroliner", "Eurofreighter", "Asialiner" and "Asiafreighter". They operated a transatlantic container service between ports on the eastern seaboard of the United States and ports in north west Europe. Following the dramatic OPEC price increases of the mid-nineteen seventies, operations were constrained by rising fuel costs. Some modification of the engine systems on those ships was undertaken to permit the burning of a lower grade of fuel (i.e. marine diesel). The modifications were partially successful. It was proved that particular fuel could be used in a marine gas turbine but, savings made were less than anticipated due to increased maintenance requirements. After 1982 the ships were sold, then re-engined with more economical diesel engines. Because the new engines were much larger, there was a consequential loss of some cargo space.

The first passenger ferry to use a gas turbine was the GTS *Finnjet*, built in 1977 and powered with two Pratt & Whitney FT 4C-1 DLF turbines, generating 55000 kW and propelling the ship to a speed of 31 knots. However, the Finnjet also illustrated the shortcomings of gas turbine propulsion in commercial craft, as high fuel prices made operating her unprofitable. After just four years of service additional diesel engines were installed on the ship to allow less costly operations during off-season. Another example of commercial usage of gas turbines in a passenger ship are Stena Line's HSS class fastcraft ferries. HSS 1500-class *Stena Explorer*, *Stena Voyager* and *Stena Discovery* vessels use combined gas and gas (COGAG) setups of twin GE LM2500 plus GE LM1600 power for a total of 68,000 kW. The slightly smaller HSS 900-class *Stena Charisma*, uses twin ABB–STAL GT35 turbines rated at 34,000 kW gross. The *Stena Discovery* was withdrawn from service in 2007, another victim of too high fuel costs.

In July 2000, the Millennium became the first cruise ship to be propelled by gas turbines, in a Combined Gas and Steam Turbine configuration. The RMS Queen Mary 2 uses a Combined Diesel and Gas Turbine configuration.

ADVANCES IN TECHNOLOGY

Gas turbine technology has steadily advanced since its inception and continues to evolve; research is active in producing ever smaller gas turbines. Computer design, specifically CFD and finite element analysis along with material advances, has allowed higher compression ratios and temperatures, more efficient combustion and better cooling of engine parts. On the emissions side, the challenge in technology is increasing turbine inlet temperature while reducing peak flame temperature to achieve lower NOx emissions to cope with the latest regulations. Additionally, compliant foil bearings were commercially introduced to gas turbines in the 1990s. They can withstand over a hundred thousand start/stop cycles and eliminated the need for an oil system.

On another front, microelectronics and power switching technology have enabled commercially viable micro turbines for distributed and vehicle power.

ADVANTAGES AND DISADVANTAGES OF GAS TURBINE ENGINES

Advantages of Gas Turbine Engines

- Very high power-to-weight ratio, compared to reciprocating engines;
- Smaller than most reciprocating engines of the same power rating.
- Moves in one direction only, with far less vibration than a reciprocating engine.
- Fewer moving parts than reciprocating engines.

- Low operating pressures.
- High operation speeds.
- Low lubricating oil cost and consumption.

Disadvantages of Gas Turbine Engines

- Cost is much greater than for a similar-sized reciprocating engine since the materials must be stronger and more heat resistant. Machining operations are also more complex;
- Usually less efficient than reciprocating engines, especially at idle.
- Delayed response to changes in power settings.

These disadvantages explain why road vehicles, which are smaller, cheaper and follow a less regular pattern of use than tanks, helicopters, large boats and so on, do not use gas turbine engines, regardless of the size and power advantages imminently available.

Diesel Generator

A diesel generator is the combination of a diesel engine with an electrical generator (often called an alternator) to generate electric energy. Diesel generating sets are used in places without connection to the power grid or as emergency power-supply if the grid fails. Small portable diesel generators range from about 1kVA to 10kVA may be used as power supplies on construction sites, or as auxiliary power for vehicles such as mobile homes.

Diesel Generator Set

The packaged combination of a diesel engine, a generator and various ancillary devices such as base, canopy, sound attenuation, control systems, circuit breakers, jacket water heaters, starting systems etc, is referred to as a generating set or a gen set for short.

While the larger industrial generators can range from 8kVA - 30kVA for homes, small shops & offices up to 2000kVA used for large office complexes, factories. A 2000 kVA set can be housed in a 40ft ISO container and be fully packaged and portable. Sizes up to about 5 MW are used for small power stations and these may use from one to 20 units. In these larger sizes the engine and generator are brought to site separately and assembled along with ancillary equipment

Diesel generators, sometimes as small as 250 kVa are widely used not only for emergency power, but also many have a secondary function of feeding power to utility grids either during peak periods, or periods when there is a shortage of large power generators.

Ships often also employ diesel generators, sometimes not only to provide auxiliary power for lights, fans, and winches, etc. but also for main propulsion. With electric propulsion the generators can be placed in a convenient position, to allow more cargo to be carried. Electric drives for ships were developed prior to WW I. Electric drives were specified in many warships built during WW II because manufacturing capacity for large reduction gears was in short supply, compared to capacity for manufacture of electrical equipment. Such a diesel-electric arrangement is also used in some very large land vehicles.

Generating sets are selected based on the load they are intended to supply power for, taking into account the type of load, ie emergency or for continuous power, and the size of the load, and size of any motors to be started which is normally the critical parameter.

POWER PLANTS - ELECTRICAL "ISLAND" MODE

One or more diesel generators operating without a connection to an electrical grid are operating in "island" mode. Several parallel generators provides the advantages of redundancy and better efficiency at part loads. An island power plant intended for primary power source of an isolated community will often have at least three diesel generators, any two of which are rated to carry the required load. Groups of up to 20 are not uncommon.

Generators can be electrically connected together through the process of synchronization. Synchronization involves matching voltage, frequency and phase before connecting the generator to a live bus-bar. Failure to synchronize before connection could cause a high current short-circuit or wear and tear on the generator and/or its switchgear. The synchronization process can be done automatically by an auto-synchronizer module. The auto-synchronizer will read the voltage, frequency and phase parameters from the generator and bus-bar voltages, while regulating the speed through the engine governor or ECU (Engine Control Module). Typical manufacturers are ComAp, GAC, Woodward and Heinzman who dominate this market

Load can be shared among parallel running generators through load sharing. Like auto-synchronization, load sharing can be automated by using a load sharing module. The load sharing module will measure the load and frequency at the generator, while it constantly adjusts the engine fuel control to shift load to and from the remaining power sources. As the prime mover of a diesel generator runs at constant speed, it will take more load when the fuel supply to its combustion system is increased, while load is released if fuel supply is decreased.

Bibliography

Barnett, P.G.: *Survey of Research on the Impacts of Pesticides on Agricultural Workers and the Rural Environment*, Davis, CA: California Institute of Rural Studies, 1989.

Bryant, Bunyan: *Environmental Justice: Issues, Policies, and Solutions*, Washington, DC, Island Press, 1995.

Burch, William R. Jr.: *Human Ecology: An Environmental Approach*, Belmont, Duxbury Press, 1994.

Charles E. Faupel: *Environmentalism and Civil Rights in Sumter County*, Colorado, Westview Press, 1992.

Daubenmire, R. F.: *Plants and Environment*, New York, Wiley, 1947.

Erickson, Kai: *A New Species of Trouble: Explorations in Disaster, Trauma, and Community*, New York, W.W. Norton, 1994.

Gay, Kathlyn: *Saving the Environment: Debating the Costs*, New York, Franklin Watts, 1996.

Gedicks, Al.: *The New Resource Wars: Native and Environmental Struggles Against Multinational Corporations*, Boston, South End Press, 1993.

Gelobter, Michael: *The Distribution of Air Pollution by Income and Race*, Berkeley, University of California, 1989.

Gerrard, Michael B.: *Whose Backyard, Whose Risk: Fear and Fairness in Toxic and Nuclear Waste Siting*, Boston, MIT Press, 1994.

Harvey L. Molotoch: *Urban Fortunes: The Political Economy of Place*, Berkeley, University of California Press, 1987.

Hill, David: *The Quality of Life in America; Pollution, Poverty, Power, and Fear*, New York, Holt, Rinehart and Winston, 1973.

Hofrichter, Richard: *The Theory and Practice of Environmental Justice*, Philadelphia, New Society Publishers, 1993.

Jaimes, Annette: *The State of Native America: Genocide, Colonization, and Resistance*, Boston, South End Press, 1992.

John M. Seidl: *Politics American Style: Race, Environment and Central Cities*, New York, Prentice-Hall, 1972.

Johnson, Rebecca L.: *The Greenhouse Effect: Life on a Warmer Planet*, Minn., Lerner Publications, 1993.

Kruvant, W.J.: *People, Energy, and Pollution*, Cambridge, Ballinger Publishing, 1975.

LaBalme, J.: *A Road to Walk, A Struggle for Environmental Justice*, Durham, Regulator Press, 1987.

Leonard G. Ritt: *Environment, Energy, and Black Americans*, Beverly Hills and London, Sage Publications, 1976.

Mandelker, D.R.: *Environment and Equity: A Regulatory Challenge*, New York, McGraw Hill, 1981.

Merchant, Carolyn: *Radical Ecology: The Search for a Livable World*, New York, Routledge, 1992.

Middleton, Nick: *Atlas of Environmental Issues*, New York, Facts on File, 1989.

Paul Mohai: *Race and the Incidence of Environmental Hazards: A Time for Discourse*, Boulder, Westview Press, 1992.

Sachs, Aaron: *Eco-Justice: Linking Human Rights and the Environment*, Washington, World Watch Institute, 1995.

Strahler, A. H.: *Introduction to Environmental Science: Santa Barbara*, New York, Hamilton Publishing Company; 1974.

Vallette, J.: *The International Trade in Wastes: A Greenpeace Inventory*, Washington, Greenpeace, 1989.

Wade, J. L.: *Society and Environment: The Coming Collision*, Boston, Allyn and Bacon, 1972.

Wenz, Peter S.: *Environmental Justice*, Albany, State University of New York Press, 1988.

Zupan, J.M.: *The Distribution of Air Quality in the New York Region*, Baltimore, Johns Hopkins University Press, 1973.

Index

□□□